HYPHOLOGY

Jake Nabasny

This text was originally composed in Spring 2012 in fulfillment of the requirements for a B.A. in Philosophy and was granted the mark of high distinction. This revision from 2015 contains some new and substantially revised sections.

ISBN: 978-1-79488-961-3
Typeset in LATEX

More information can be found at:
http://jake.nabasny.com

Buffalo, NY
2020

To Ashley
for her unwavering
support and love

CONTENTS

1

Introduction

Writers, like people, have defining moments. These moments are sometimes subtle turning points, gathering points, or even tipping points on the verge of radical change. Sometimes we can even locate these sensitive points within a writer's work. In a non-reductive manner, the critic can begin to speak about a kernel or center of an author's life and work. This task can be non-reductive because one is never able to count the multiplicity of divergent paths that stem from this center. Philosophers are not immune to this structure. There are even some who explicitly point to their center.

For example, George Berkeley wagered his entire philosophy on one indubitable proposition. The center poses an issue for any writer since it both produces the space of their thought, but limits it at the same time. In this way, philosophies are nebulae that sparkle under the tremendous force of fusion, but burn out once they are stretched too thin. The development of structuralism was a response to this problem of the center.

For all of his dispersion of meanings, eye winks, sighs, and erotic maxims, Roland Barthes' structuralism admitted a center of its own. When discussing the representation of Romans in film, Barthes writes about the various hair styles which the film *Julius Caesar* seems to revolve around. He announces rather early in the essay: "the bald are not admitted" (*les chauves ne sont pas admis*).[1] The meaning of "admit" (*admis*) is crucial here since it carries the same double meaning in French. In admitting, one both allows entrance into a space and gives a confession; the very division between internal and external is abolished. This is precisely the action that Barthes negates in his structuralism; not admitting an exterior becomes the center of his structuralism. By not admitting the Romans that are bald, structuralism is effectively enclosed and the boundary of every myth is clear and distinct: only Romans with hair are significant. One finds the alternative to this closure in Søren Kierkegaard who invokes the myth of Proserpine to in-

---

1. Roland Barthes, *Mythologies*, trans. Annette Lavers (New York, NY: Hill and Wang, 1972), 26.

clude even the bald in the act of plucking a hair.[2] It is around this question of baldness that the entire battle between Barthes' structuralism and Jacques Derrida's deconstruction will be fought. After *Of Grammatology* was published in 1967, critical attention shifted from the center of the structure to the center of structuralism itself.

The first question of this investigation will be: What is structuralism? Structuralism itself has taken many forms and in each instance seems to be concerned with a different goal. For the duration of this book, I will adopt a generalized structuralism that is taken up by Barthes, criticized by Derrida, and is largely inherited from Ferdinand de Saussure. Barthes acknowledges his debt to Saussurean structuralism, but also extends its objects of analysis further. Derrida, on the other hand, is primarily criticizing Saussurean structuralism, but also addresses other structural linguists like Louis Hjelmslev. Therefore, it is a method derived from Saussure and Barthes that I will refer to as "structuralism."[3]

The unique instrument developed to breakdown structures is called the "sign." The sign is distinguished from its referent, which is the object in the world. In traditional structuralism (e.g. linguistic), the sign is equated to the amalgamation of a word and its meaning. Insofar as the sign is divided into two distinct parts,

2. "No one escaped my attention. Like Proserpine, I plucked a hair from every head, even the bald ones." Kierkegaard is referring to the myth that Proserpine would pluck a hair from every person who died before they could enter the underworld. One assumes that even the bald must somehow be admitted since this myth does not exclude the possibility of a bald person existing in the underworld. Søren Kierkegaard, *Repetition*, trans. Howard V. Hong and Edna H. Hong (Princeton, NJ: Princeton University Press, 1983), 170.

3. Sometimes I will use the word "semiology" or "semiotics" to denote the same method. Saussure first proposed the word "semiology" as an extension of his linguistic system to other systems. Barthes then appropriated this term when he began to structurally analyze the garment industry.

these components are referred to as the signifier and signified, respectively. The signifier is the phonic or graphic aspect of the word ("cat"); while the signified is the concept it represents (a domesticated feline). It is relevant to note that Saussure saw this distinction as purely abstract for the sake of clarification; these distinctive parts are like the two sides of a single sheet of paper.[4]

"Signification" is the process by which a sign's meaning is selected. At the most basic level, signification draws a link between a signifier and a signified in order from them to function as a sign. However, the sign's meaning does not only rest on the correlated signifier and signified, but also on how the sign differs from other signs around it. For example, the sign "bark" will have a different meaning depending on the word that precedes it. "Tree bark" points to a different signified than "dog bark." In this way, structuralism posits that the identity of terms consists in their differences from other terms in the system; they have no positive meaning independent of the system. More will be said on this below.

Structuralism has two goals: (1) to determine the mode of meaning-production and (2) to be able to make meaningful statements about signs. The revolutionary insight in the first goal is that meaning does not merely exist, but is produced. The assumption that a word has an innate meaning cannot account for the transformations that meaning undergoes over time. While structuralists are usually opposed to analyzing structures over time, they acknowledge that this change occurs. Each structure produces meaning in a different way. The structuralist's task is to determine how an ut-

---

4. The sheet of paper analogy is quite famous in scholarship on Saussure and is even taken up by Derrida. For the original remark, see Roland Barthes, *Course in General Linguistics*, ed. Charles Bally and Albert Sechehaye, trans. Roy Harris (Chicago, IL: Open Court, 2006), 111.

terance gains its meaning. For example, "the cat is on the mat" is an empty string of signs until one knows the rules that are operating on them. The rule or law of a structure (e.g. what grammar is to language) sets the proper ways in which signs can be configured to produce meaning. The ultimate purpose of structuralism is to find these hidden rules.

The second goal that leads to realizing these rules is to understand signs. Unlike the grammar of a structure that determines which utterances are well-formed and meaningful, a sign is also subject to its own categories which determine the meaning of that individual sign. The structuralist must figure out which signs can be substituted for each other along with their appropriate signifieds. This order of substitution will come to be known as the paradigmatic relation between signs. Such a possibility is a central aspect of structuralism, in Barthes especially, since a ritual in a wrestling match poses more interpretive difficulty than a word in a sentence.

To accomplish its two goals, structuralism must clarify three components of every structure: the total structure, the chains of individual signs that can possibly be produced, and articulation. The total structure helps clarify the boundaries of any given structure. The determination of boundaries incites the closure of structuralism hinted at above. For instance, to analyze the varying hair styles in Roman films, Barthes sets a limit in saying that the bald are not admitted. The second and third components relate directly to the two goals stated above. The possible chains of signs that can be produced are governed by the grammar; "articulation" is the various ways a sign is positioned within its context that give it its meaning. For example, the word bear can be articulated in two very different ways: (1) "She could not *bear* the pain," and

(2) "Look out for the *bear!*" Grammar, as it defines the differential relations between signs, aides in clarifying which signified is
attached to "bear." Once these aspects are understood, the structuralist can make judgments concerning possible future utterances
within that structure. Some structuralists even attempt to dis-
cover the parameters for what constitutes a well-formed utterance.
Therefore, structuralism, in its most rigorous executions, is a science of meaning. It has defined a general methodology complete
with fundamental concepts. In this way, it should be distinguished
from literary or political modes of analysis that often read the same
meaning into every structure (e.g. Marxist criticism).[5]

In *Of Grammatology*, Derrida forms his most explicit critique
of Saussurean structuralism. Along with structuralism, he is also
targeting the other popular philosophies such as phenomenology
and hermeneutics. One of the most popular misconceptions about
Derrida (and there are quite a few) is that he defends a relativist
position in which any attempt at an objective analysis of the world is
entirely meaningless. The intention of his method of reading, which
he calls "deconstruction," is not to defend a position of his own,
but to reveal internal inconsistencies of the philosophies that claim
to make clear and present descriptions of the world. Sometimes
this method involves dissolving apparatuses of meaning-production
down to their bare dichotomous relations in order to show that
the distinctions being made are not as clear as they first seem (e.g.
signifier-signified). Barthes will eventually turn to a similar skeptical
view of structuralism, especially in *Empire of Signs* and parts of

5. This is not to say that Marxist themes are entirely absent from structuralism. Some of Barthes' most enlightening insights in Mythologies concern class
struggle and the myths surrounding the middle-class.

S/Z. However, the structuralism practiced by the early Barthes entertains many concepts that are absent in Derrida's treatment.

The goal of this book is to answer Derrida's critiques of structuralism. In doing so, I will formulate a new kind of analysis that I will call "hyphology." This task will require a reconceptualization of the role of repetition in structuralism. The point is to acknowledge that some of Derrida's arguments are invidious to a closed form of structuralism, but can be answered with a few adjustments. In the first section, I will explicate Derrida's arguments against structuralism from *Of Grammatology* and *Writing and Difference*. The second section will be a comprehensive discussion of structuralism as it is practiced in Barthes. I intend to discuss the concepts he formulates that act as wedges in structuralism, pushing it to a point that Derrida did not consider. The third section offers a new approach to repetition as it is conceived by Gilles Deleuze and his precursors. They show how various features of experience can lead to an engaged structuralism. The fourth section will be an attempt to synthesize everything discussed previously. Hyphology will be elucidated in this part by combining the new formulation of repetition with the wedges from Barthes. It will then be weighed against Derrida to show where it surpasses ordinary structuralism as a method of analysis. The fifth and final section will be a series of hyphologies that demonstrate hyphology in practice.

2

Derrida and the Closure of
Structuralism

The strategy with which Derrida approaches structuralism is a concise examination of Saussure that seeks to reveal internal inconsistencies, especially with regard to meaning-production.[1] A key concern for structuralism is isolating a structure enough to be able to

1. Meaning-production is a key target of Derrida's critique. For example: "To be a structuralist is first to concentrate on the organization of meaning." Jacques Derrida, *Writing and Difference*, trans. Alan Bass (Chicago: University of Chicago Press, 1978), 26.

investigate it as objectively as possible. Saussure found it best to view language at a certain point in time rather than tracking it across its historical development, which was the dominant method at the time. He was concerned with the synchronic elements (stable and atemporal) of the structure, whereas the diachronic elements (dynamic and historical) created too much variability. Focusing on the synchronic elements provided a stronger base for analysis and gave credit to structuralism's aspirations to be scientific. For Derrida, this act of closing off a structure in order to establish a scientific method will become problematic. As I will show, diachrony is at the heart of Derrida's critiques, not as a reversal of Saussurean dependencies, but as a spotlight cast on the inescapable passage of time.[2] Every operation (trace, differance, spacing, etc.) that is developed in reaction to structuralism has time as its most basic principle.

Although a kaleidoscope of other problems are attributed to Saussure, Derrida clearly finds some more invidious than others. The beginning of his critique, which is primarily situated in *Of Grammatology*, takes a kind of distance, which is spatial and temporal, to be the opening for a critical discussion of Western metaphysics. This discussion would appear in nearly all of Derrida's works either implicitly or explicitly. The chronology of Derrida's works becomes important at this point. The year of 1967 could certainly be called an event, with all the weight this term holds today. Three dense books were published by Derrida at this time: a commentary on Husserl, a collection of essays, and a lengthy philosophical treatise on Western metaphysics. The book on Husserl, *Speech and Phenomena*, contains a critique of the subject, but

2. Derrida, *Writing and Difference*, 14.

Derrida   11

does not deal explicitly with structuralism in the way that the other
two books do.[3] It is with Husserl that *Of Grammatology* begins,
whether we see his name explicitly cited or not. He plays a promi-
nent role in all of Derrida's philosophy around this period, since he
also appears in the third book, *Writing and Difference*. A reaction
to the Husserlian notion of self-presence is what gets the ball rolling
in *Of Grammatology*.

Husserl's system requires that the subject be absolutely present
to herself. Truth can only be grounded if there is a Cartesian
self-certainty which verifies all of an individual's beliefs. Only
through the correspondence involved in representation can the subject identify a faithful relation between two terms (these terms
can take the form of, e.g., word and referent, subject and object, intelligible and sensible, etc.).[4] Derrida boils down Husserl's
formulation of the self-certainty of self-presence to the following:
hearing(understanding)-oneself-speak.

The subject, in hearing(understanding)-oneself-speak, produces
representations. These representations, which are simultaneously

3. Jacques Derrida, *Positions*, trans. Alan Bass (Chicago: University of
Chicago Press, 1981), Derrida has commented in an interview that *Of Grammatology* should be read in the middle of *Writing and Difference*, and vice
versa. In this way, these two texts form an intimate relation which excludes
Speech and Phenomena.
4. This understanding of representation is also implicated in Nietzsche
when he discusses the social development of language and consciousness (Cf.
Friedrich Nietzsche, *The Gay Science*, §354). This correlation to Nietzsche is
important to establish at the outset of our investigation into Derrida because
we must remember at all times that deconstruction takes the entirety of Western metaphysics as its target. Positive remarks about Nietzsche, Heidegger,
and Freud are lightly sprinkled around *Of Grammatology*, but, despite their
accomplishments, Derrida believes that they had failed in their pursuit of a
post- or anti-metaphysical philosophy. Even Saussure should be commended
for his attempt to construct a system of language based wholly on differences.
What Derrida will later call "grammatology" is largely his effort to carry these
philosophies to their intended conclusions.

produced and apprehended by the subject, are constituted by the fundamental structure of subjectivity. When I speak, I also hear myself speak. In speaking, I also understand. If I misspeak, I am able to immediately correct myself. Therefore, speech is always split into two parts: speaking and hearing. Thought itself does not escape this profound doubling. To think is to represent to oneself an idea in language. I can hear(understand)-myself-speak without ever uttering a word. In sum, these are the vital features of representation for Derrida: mediating, doubling, repeating, and reflecting.[5] These features pose a problem for the unperturbed self-presence that the Husserlian ego relies on. This problem pertains to structuralism just as much as it does to Husserl's phenomenology.

   Saussure is insistent that the division of the sign into the signifier and signified is only psychological.[6] Yet he also maintains that the signifier and signified evolve at different rates which suggests an unbridgeable dissimilarity.[7] In effect, a signifier must catch up to its signified. Even if he is only concerned with synchronic elements, the adhesive between the two components of the sign is loosened and the salary of a Swiss linguistics professor cannot buy enough tape to hold the pieces together. This is only a cursory remark on an argument Derrida will later tease out, but already one is able to see how the problem of representation is grafted onto the signifier-signified distinction.

---

   5. "Representation mingles with what it represents, to the point where one speaks as one writes, one thinks as if the represented were nothing more than the shadow or reflection of the representer. A dangerous promiscuity and a nefarious complicity between the reflection and the reflected which lets itself be seduced narcissistically. In this play of representation, the point of origin becomes ungraspable." Jacques Derrida, *Of Grammatology*, trans. Gayatari Chakravorty Spivak (Baltimore: Johns Hopkins University Press, 1997), 36.
   6. Barthes, *Course*, 66.
   7. Ibid., 74–8.

The sign is necessarily doubled into signifier and signified: the subject speaks-signifies to himself. Separation, difference, and delay are necessary components of the sign. Trying to add the origin to any representation, or the thing to its image, exemplifies what Derrida calls the *law of addition*: one plus one makes three.[8] Here Derrida is speaking of the referent of the sign, which is the material object that is represented by the sign (e.g. the tree as bark, wood, and leaves). We often believe that words refer directly to objects, but they are really pieces of a structure in which each word consists of two parts: signifier and signified. So, if I add two terms together (sign + referent), I actually end up with three terms (signifier, signified, and referent). Due to this outcome, concepts like the subject or the sign are revealed to be more complicated than initially supposed; each word is a multiplicity even though it appears to be a singularity. It becomes impossible to learn how to count. The structuralist can no longer count the signs within the system; the phenomenologist can no longer count the ways she loves thee. *Every number line should begin at two.*

The word "every" should not be negated as mere hyperbole. When Derrida analyzes the role that representation plays in Western metaphysics, he is calling attention to what he believes to be an historical tend (an *episteme* or paradigm) that dates back to ancient Greece.[9] His key claim is that Western metaphysics always has adopted the same assumptions and that these assumptions are inextricable from their milieu (which accounts for the failures of Heidegger and Nietzsche).[10] Derrida does not give an adequate de-

8. Derrida, *Of Grammatology*, 36.
9. Derrida, *Writing and Difference*, 3-4 and 278.
10. Ibid., 280–1.

fense of this claim in *Of Grammatology*, but is merely setting up his argument that the concept of the sign is universal even if it goes by different names. Nonetheless, the signifier-signified relationship already has shown a family resemblance to phenomenological representation.[11]

Derrida argues that representation necessitates a split between image and meaning, which suggests that meaning is never immediately present. So, the role of the absolutely present subject is grounded within a philosophy of presence. The systemic privileging of presence as somehow inherently more meaningful is what Derrida hopes to overcome. The subject is always already connected to various conscious faculties (e.g. memory, perception, imagination), which themselves are inextricable from presence. These faculties, according to Derrida, are concerned with what is present because they are directed at clear and distinct objects, such as what is remembered or what is consciously perceived. He sees this as an overt negation of the absent factors that co-constitute reality. There is a white noise behind every perception; there are echoes that irredeemably color the object that is perceived.[12]

It is no longer possible to support the self-present self-certainty of the Husserlian subject: every invocation of this subject is stillborn. Since Derrida's argument is against a certain kind of subject, the Husserlian subject that is absolutely self-present, it cannot be assumed to be an indictment of all subjectivity. However, it is through this critique that Derrida aims to deconstruct what he

---

11. For a similar reading of Lévi-Strauss' intelligible-sensible distinction: Derrida, *Writing and Difference*, 281–9.

12. Husserl's notion of horizon attempts to deal with these absent factors, but merely posits that they exist and not, as Derrida claims, that they have a direct effect on the faculties. Leibniz's "little perceptions" are similarly shortcoming since they only include the supersensible and not the non-sensible.

believes to be the foundations of Western metaphysics, which worships the sign as its logic and expression. This critique begins by demonstrating that the charge of meaning produced between the signifier and signified has been explicated in at least two fundamental ways. The difference between the two hinges on, what Derrida calls, the transcendental signified.

In his essay, "Structure, Sign and Play in the Discourse of the Human Sciences," Derrida proclaims structuralism to be an eventual rupture that reinterprets "structure," which has been the dominant theme of Western metaphysics. Prior to the structuralist rupture, "structure" was understood as having a center that determined its function and purpose. This center, however, Derrida points out, was never within the structure itself, but existed outside of it as a way of grounding the structure to some fundamental truth or reality. This center, which was not center, is the transcendental signified. It is "transcendental" because it is outside of the structure and "signified" because it establishes the means of meaning-production for the structure. Derrida cites a long list of examples taken from the history of philosophy: "*eidos, arche, telos, Energeia, ousia* (essence, existence, substance, subject) *aletheia,* transcendentality, consciousness, God, man, and so forth."[13]

Structuralism represents a radical break with this traditional of locating the center because it attempts to think the "structure" as a whole, rather than just its organizing principle. It is the "structurality of structure" that becomes the true object of structuralism, not only as a possible object of thought, but also of repetition.[14]

---

13. Derrida, *Writing and Difference*, 279–80.
14. Ibid., 280. Notice how Derrida appends the issue of "repetition" to the structuralist rupture. This theme will come into focus more clearly below.

Derrida's description of this moment implicitly recalls Saussure's turn away from naturalism. Instead of assuming a natural relation between signifier and signified that was established by a transcen- dental signified (or "center"), Saussure argued that the relationship is arbitrary. There is thus no center that fixes the position or rela- tion of signs. Derrida's general description of this disruption goes as follows:

> Henceforth, it was necessary to begin think- ing that there was no center, that the center could not be thought in the form of a present- being, that the center had no natural site, that it was not a fixed locus but a function, a sort of nonlocus in which an infinite number of sign- substitutions came into play.[15]

A theme hinted at in this passage that will reoccur several times is the theme of infinite substitution. Derrida's challenge to struc- turalism is brilliant because he does not tactlessly break open its closure by demanding that it recognize external factors of meaning- production. Rather, his challenge tacitly assumes the closure of structures, but points to its previously unrealized consequence. Al- though the field of signs may be finite, the movement from signifier to signified engenders an infinite maneuver in the form of substitu- tion. How does this work?

The signified can only become intelligible if it is present. To this end, it must be represented in some way. Yet, any representation to a language user must take the form of a signifier. In this way, the hoped for signified with its finality and presence of meaning has slipped away. What the signifier has led to is simply another

15. Derrida, *Writing and Difference*, 280.

signifier. Representation goes on infinitely when there is not a positive, singular meaning at its core that can act like an anchor. This eternal reiteration of signifiers marks the uniqueness of the structuralist moment, which Derrida refers to as "play."[16]

Play inevitably challenges the efficacy of meaning-production. If the ultimate, present meaning of a signifier is perpetually delayed, then meaning is never produced. The structuralist promises to understand the meaning of structures in themselves, but creates a methodology that conceals meaning precisely as it claims to reveal it: "that which also metaphysically menaces every structuralism: the possibility of concealing meaning through the very act of uncovering it. *To comprehend* the structure of a becoming, the form of a force, is to lose meaning by finding it."[17] A concrete example of play would be the dictionary. A word is attached to a definition that supposedly elucidates the meaning of that word. The meaning, however, is articulated through the use of a multiplicity of other signifiers. These signifiers, in turn, would have to be investigated and comprehended. Inevitably, any search for a pure definition will only result in a pile of signifiers and no signifieds.

The dictionary example helps clarify two issues. First, the objective of the signified is to present and re-present meaning. This structuralist desire is aborted as soon as one sets off down the rabbit hole of signification. In this way, meaning is perpetually delayed and deferred to other signifiers. The second issue concerns the structuralist turn. What makes structuralism an "event" and "rupture" for Derrida is that it flouts the necessity of the tran-

---

16. He also notes that this "play" is evident in all destroyers of metaphysics (Nietzsche, Heidegger, and Freud, particularly).
17. Derrida, *Writing and Difference*, 26.

scendental signified. Traditionally, we approach a dictionary by assuming a transcendental signified which allows the dictionary to complete its primary function: defining. Structuralism isolates the text by negating the existence of such an organizing principle. This closure has the unintended consequences of the infinite substitution of signifiers and, ultimately, the impossibility of meaning.

Despite all of its problems, we need the sign. We need it in order to critique it. Derrida does not want to assess the sign by external criteria, but to reveal that it is internally inconsistent.[18] There is a radical difference within the sign between the signifier and signified. This difference is supposedly irreducible, yet it must be reduced in order for the sign to function. Sometimes Saussure will speak of the "signifier," but other times it becomes the "sign of..." So, one encounters an accordion motion oscillating from the sign to the radical difference of its components. It is here that Derrida locates an aporia. As soon as the sign is reduced to a single term, it also needs the opposition that it is reducing. The sign is destroyed as soon as the signifier is relegated to the signified because it becomes a one-dimensional positive term, which is exactly what Saussure was trying to avoid. Therefore, it is necessary that a critique of the sign appropriates the notions of signifier and signified. We cannot, as Derrida reminds us, simply forget our history as a solution to the problem of the sign.[19] What is at issue here is the existence of time. The signifier and signified cannot appear in the same moment. There needs to be a second moment in which the signified appears *after* the signifier. So, signification can never properly occur because there is never a *sign* that one can point to in a

18. Derrida, *Writing and Difference*, 281.
19. Derrida, *Of Grammatology*, 28.

single moment. Located in this concern is one of Derrida's most revealing concepts, which he calls alternatively trace and *différance*.

Before illuminating the elusive nature of meaning-production, Derrida discusses what can be called the founding principle (*arche*) of Saussurean linguistics: the arbitrariness of the sign thesis. The linguist's interest is not in saying what is necessary for structuralism to exist, but figuring out the ordinary conditions for the possibility of language, which vary depending on the complexity of the language (e.g. resemblance between signifier and signified). So, when the signifier "tree" is fixed to a signified, there is nothing in the phonic or graphic representation of the word "tree" that hints at what the signified could be. Without any connection or association between the two terms, Saussure will say their relationship is arbitrary. At the heart of this argument is Saussure's demand that all signifier-signified relations should be unmotivated. If there were a motivation for a signifier to favor a certain signified (whether this motivation is subjective or not), then the role of difference vanishes from Saussure's system. The relation must be arbitrary for substitution and comparison to occur.[20] Certainly, the arbitrariness of the sign thesis seems commonsensical.

This thesis, as we have already mentioned, is contrasted to a natural connection between the signifier and signified. Saussure admits that some natural connections exist in language, specifically onomatopoeia, but he believes these exceptions to be irrelevant.[21] "Bam!" is naturally related to its signified, but such a word is a rare and extravagant usage of language. Like David Hume's missing

20. The paradigmatic relations would vanish, leaving a sign to be a mere, positive term. This will be described in full below, but it is important to note in this discussion what is at stake.
21. Barthes, *Course*, 69.

shade of blue, Saussure will admit the problem exists, but decides not to engage with it.[22] Derrida is not concerned with this counter-example. He is more interested in the natural connections that exist outside of the signifier-signified relation. He argues that there are two natural (as opposed to arbitrary) relationships that exist prior to the arbitrariness of the sign: (1) the signifier-signified relation proper and (2) the relation of writing to speech.

Every signifier must naturally be attached to a signified. This is the primary relation that Saussure had not realized. There is always a motivation for a signifier to find a signified. Essentially, Saussure is building a puritan utopia where every man has a woman. Although it is arbitrary whether the man walks with a limp or the woman has blond hair, it is necessary that they find each other. In the second natural relationship, Derrida is concerned with Saussure's attempt to chase "writing to the outer darkness of language."[23] There is a motivated link from writing to speech in that the grapheme is always a representation of the phoneme. Not only does this assert the primacy of speech over writing ("phonocentrism"), but it also reintroduces the problems of the sign that were discussed above.

The relationship between writing and speech can be understood in the same way as that of the signifier and signified. Mainly, the

---

22. The comparison between Saussure and Hume is interesting. They are both building systems based on a founding principle: the copy thesis and the arbitrariness of the sign. For Hume, the existence of an idea that is not copied from an impression, but is rather derived from previous perceptions, has the same externality as a signifier with a natural connection to its signified. The relationship between language and metaphysics is paramount for Derrida: "...such a linguistics, whether spontaneous or systematic, has always had to share the presuppositions of metaphysics. The two operate on the same grounds." Derrida, *Of Grammatology*, 21.
23. Ibid., 45.

grapheme cannot be an "image" of the phoneme because it must always already be reduced into a single unit (e.g. the sign). However, there is an insistence in Saussure that speech should be privileged over writing at all times.[24] The claim is that speech inherently holds more clarity, presence, and meaning than writing, which is absent, distracted, and partial. It is believed that the sound-image is more closely related to the signified than the grapheme; the voice is thought to be consciousness itself, and the signifieds are nearest to me when I speak. When the signifier asks "where is it?" the signified answers "here it is!" Derrida has shown that language, as conceived by the structuralist, does not facilitate the discovery of meaning, but always the endless search for it. Every sign says: "Where?" The signs accumulate and begin to resonate; one is able to hear a faint voice in the background of every utterance that says, "Not here." This adventure of meaning is always emphin search of lost time: the time that is required to follow every signifier to its desired end, even though the end does not exist. Searching and locating eternally–these are the characteristics of the trace.

It is clear that the trace, as the name implies, is a movement. It is a becoming that constitutes signs and entities in general. Derrida argues that there must be an unmotivated force behind signification that remains faithful to Saussure's schematic and preserves difference as the crucial principle of structuralism. As we have seen so far, the rupture of structuralism inaugurates the infinite play of signifiers. The ever evasive signified allows for an endless tracing of signifiers. But if the trace never ends up tracking something down, why is it important? The trace is important not because of where it leads, but in how it leads. It composes each sign as it moves

24. Barthes, *Course*, 29 and 111.

through every structural possibility by constant substitutions. Just like the signifier-signified relation, the trace is unmotivated, but in a way that must be distinguished from arbitrariness, naturalism, and subjective preference. We have already noted how the arbitrariness thesis was dubious. Alternatively, Derrida is not asserting a natural connection beyond the arbitrariness of the sign; he has shown Saussure's own view to be inconsistent and is attempting to explain the process gestured at by structuralism. Furthermore, Derrida agrees with Saussure that the signified of a signifier cannot merely be the choice of a speaker. There is a socio-cultural sphere in which a certain signified is considered acceptable for a given signifier. The speaker necessarily inhabits this sphere and would not be using language (either properly or at all) if she decides her own meanings for the words, although speakers can sometimes suggest changes of word meanings when a unique use of the word is signaled (e.g. slang).

Thus, it is clear what the trace is not. What is it exactly, though? This question cannot be asked of the trace in any practical way. Any existential question implies articulation, that something is presently manifested in a determined way. Since the trace is the movement that composes articulation, the becoming that crystallizes into being, it is not an articulation itself. It would be impossible to derive the becoming of being from being itself just as it would be problematic to reduce force to form. Yet this becoming certainly exists as a process of determination. At this point, it can be argued that Derrida is being unclear or intentionally obscure. This accusation is far from the case. It is not that Derrida is being unclear, but that he is discussing the very conditions for clarity, which themselves must precede clarity.

As a condition for the intelligibility of all signifiers, the trace is erased in its own becoming.[25] The formulation of the trace is similar to the theorization of gravity that is derived from the fable about Newton and the apple. Like gravity, the trace is not frozen at a certain point or velocity. In the very same way in which gravity is acceleration, the trace is becoming. What does it become? It becomes unmotivated in-itself. It is at the core of both genetic and structural possibilities. These possibilities determine the field of presence that the sign eventually inhabits. (When I say possibility I mean the space of difference available to any sign. This possibility is also called substitution.) So, the trace does not refer back to any present nature or origin, it is "indefinitely its own becoming-unmotivated."[26] It is the groundlessness that all of Western metaphysics is founded on and engenders the movement from signifier to signified, possibility to reality.

Derrida creates a neologism to talk about the existence of the trace and the radical difference between the signifier and signified: *différance*. The word itself entails all the play and ambiguity that Derrida finds in language. It is the same word in French for "difference" (*différance*), except one letter is changed. The neologism is, therefore, pronounced the same as "difference," but is written differently. Already one finds a transition to thinking about writing as equal to speech in degree of presence. This difference can

---

25. Derrida, *Of Grammatology*, 47.
26. Ibid.

only become present in its written form.[27]   The meaning of the
neologism is in its very ambiguity. *Différance* means both to differ
and to defer. *Différance* (with an *a*) engenders and unites both
definitions.[28] The distance or difference foregrounded by *différance*
is of a temporal and spatial character. Derrida highlights this two-
dimensionality in his essay, "Différance."

Différance* is introduced as "literally neither a word nor a concept."[29] Just like the trace, it is difficult to define or pin down in
any clear and distinct fashion. Derrida partly recalls his critique of
the concept of the sign as a gesture toward the delay and displace-
ment that is constitutive of *différance*. For example, the signifier
cat differs from its signified; the meaning of the sign "cat" is constituted by the trace from its signifier to its signified. Therefore,
the meaning (signified) of cat is always differing and deferring from
its image (signifier). This causes a delay in signification, a delay
which lasts indefinitely because, as we have seen, there is no final
signified. Thus, *différance* can be understood as that difference,
internal to the sign, which allows for the existence of the play of
signs (which is itself a secondary kind of difference).[30]

Great pains are taken to distinguish these two kinds of differ-

27. It is vital to make a distinction concerning this issue in Derrida. He is
adamant about not reversing the power dynamic between writing and speech;
he does not want to make writing the new, self-present meaningfulness that
speech engenders. However, he does speak about an *arche-writing* that comes
before both writing and speech. This will not be discussed in this essay because
it falls outside of the focus on structuralism, but it is important to remember
that Derrida does not want to assert any term over another. He simply wants
to show how what we usually believe to be polar opposites often overlap,
penetrate, and envelop each other.
28. I have maintained the convention of keeping *différance* untranslated.
29. Jacques Derrida, *Margins of Philosophy*, trans. Alan Bass (Chicago: University of Chicago Press, 1982), 3.
30. Ibid., 9.

ence. Derrida spends the majority of the essay linking *différance* to similar theoretical notions in Hegel, Heidegger, Nietzsche, and others. Unlike these other theorists, Derrida questions the extent to which *différance* could be considered an "origin" or producer of secondary difference. Although it "produces" the difference between signifier and signified, speech and writing, and so on, it is only in the form of a negative theology that *différance* can be encountered.[31] In the same way, it would be a mistake to conceive of *différance* as an activity, since it is "no more static than it is genetic, no more structural than historical."[32] *Différance*, then, is the synthesis of these binary opposites in their very impossibility of synthesis, much like the presence of the sign is predicated upon the *différance* of its components.[33]

Indeed, the presence of the sign can only come to us through the differences (i.e. absences) that constitute it. Analogically, Derrida argues that phonetic writing requires nonphonetic signs (e.g. punctuation) in order to function. These other elements founded a "silent play" between the audible phonemes and the inaudible signs that separate them: "Inaudible is the difference between two phonemes which alone permits them to be and to operate as such. The inaudible opens up the apprehension of two present phonemes such as they present themselves."[34] This detour through the theme of silence is meant to highlight the nonphonetic difference between

31. Ibid., 6.
32. Ibid., 12. Derrida makes a similar argument about the trace in *Of Grammatology*, 47–8.
33. "And it is this constitution of the present, as an 'oringinary' and irreducibly nonsimple (and therefore, *stricto sensu* nonoriginary) synthesis of marks, or traces of retentions and protentions, [...] that I propose to call archi-writing, archi-trace, or *différance.*" ibid., 13.
34. Ibid., 5.

différence and *différance* (the *e* and the *a*) since these words are pronounced the same. Since phonetic writing must also rely on non-phonetic signs to be what it is, Derrida contends that there is no "purely and rigorously phonetic writing."[35] For the same reasons, an inaudible difference must exist between graphemes in graphic writing. Hence, the difference between the e and the a points to an order that resists articulation in sensibility and intelligibility, vision and sound.

It is easy to misconstrue what Derrida is suggesting with his comments on the inaudible difference of *différance*. Russell Daylight, in an unambiguous defense of Saussure, takes issue with Derrida on this exact point. *What if Derrida was wrong about Saussure?*, an aptly titled text with a curious excess of block quotes, lays it all on the line against Derrida: "difference *is* audible."[36] His seemingly knockout example (since it is the only one) comes in the form of Morse code. In Morse code, Daylight argues, the silence between phonemes is itself a phoneme. That is to say, the silence serves some signifying function and can only do so insofar as it is heard. Yet this example entirely misses the point. *Différance* is not a relative difference between phonemes, but a radical difference encoded in any structure. It is the difference (as delay) that allows a syntagmatic chain to be constituted. Without this delay, or "silence," the sign could not exist as such. In the case of Morse code, the durations of soundlessness vary according to different possible significations (e.g. new letter or new word). These durations come to a close and take on their signifying force by the manifesta-

---

35. Derrida, *Margins*, 5.
36. Russell Daylight, *What if Derrida was Wrong about Saussure?* (Edinburgh: Edinburgh University Press, 2011), 119.

tion of the next signifier. It is the gap between the pause and the beep that cannot be transcribed and what constitutes phonemes as such. This silent gap is constitutive for signification, but, as Derrida rightly suggests, it is also signification's unraveling.

This point is especially clear when Derrida discusses various manifestations of *différance*.[37] One that is particularly close to the current topic is spacing.[38] The discussion of spacing in *Of Grammatology* evolves out of a concern for the "pause, blank, punctuation, interval in general, etc." that interrupts the chain of signs in writing (i.e. the nonphonetic signs mentioned above).[39] These "spaces" create the possibility for signification within their own erasure.[40] There must be a distinction between a signifier and signified for there to be signification, yet it is this space (and time) between the two terms that makes it impossible to fold them over into one moment. Due to this irreducible space between the signifier and signified, Derrida recommends the idea of a hinge (*brisure*) to understand this paradoxical relationship.[41] Above all, Derrida asserts that spacing is purely horizontal, which means that the difference between the signifier and signified is on the same plane and cannot

37. The trace, which we have already seen, is one of these manifestations: "The (pure) trace is differance." Derrida, *Of Grammatology*, 62.

38. Space, for Derrida, implies a distance or making-distant of things and meanings: "In constituting itself, in dividing itself dynamically, this interval is what might be called spacing, the becoming-space of time or the becoming-time of space (temporization)." Derrida, *Margins*, 13.

39. Derrida, *Of Grammatology*, 63.

40. "This signification is formed only within the hollow of differance: of discontinuity and of discreteness, of the diversion and the reserve of what does not appear." ibid., 69.

41. "The hinge marks the impossibility that a sign, the unity of a signifier and a signified, be produced within the plentitude of a present and an absolute presence." ibid.

be side-stepped or justifiably hierarchicalized.[42] In this way spacing can be interpreted more literally to be the actual white spaces between words on a page, since this too constitutes a delay. One is not able to escape the breaks and gaps of writing by simply moving vertically. The philosophical concepts that have been discussed up to this point must be thought of as working in one dimension. Whether this is a limitation inherent in structuralism or one that Derrida places on it is one of the major problems I will address over the course of this book.

The general method that has been developed by Derrida and explicated above is termed "deconstruction." Derrida inherits the word from Heidegger (*destruktion*), but builds on it considerably. De(con)struction is a double gesture that attempts to invert the privileged term of metaphysical dichotomies at the same time that it seeks to displace the rigor and purity of that dichotomy:

> Deconstruction cannot limit itself or proceed immediately to a neutralization: it must, by means of a double gesture, a double science, a double writing, practice an *overturning* of the classical opposition *and* a general *displacement* of the system. [...] Deconstruction does not consist in passing from one concept to another, but in overturning and displacing a conceptual order, as well as the nonconceptual order with which the conceptual order is articulated.[43]

Like *différance*, deconstruction is a stratagem that can be deployed when one engages with texts. Its goal is not to invert the

42. "...the horizontality of spacing, which is in fact the precise dimension I have been speaking of so far..." Derrida, *Of Grammatology*, 69.
43. Derrida, *Margins*, 329. This double gesture is also described in *Writing and Difference*, 20.

dichotomies of speech/writing, presence/absence, and so on, but to undercut their supposedly self-evident justification and violent reinforcement of dominant metaphysical hierarchies. For this reason, it is no surprise that Derrida's later work took an ethical and political turn.

Before departing from Derrida, we must deal with a few of his arguments against Saussure that do not apply to Barthes' form of structuralism. It is not the case that Barthes "corrects" these issues, but is merely concerned with different structures outside of language which force him to deal with elements not entertained by Saussure. First, excluding diachronic elements from structural analysis negates the historical contingency in which each sign is formed. Saussure realized that the meanings of words are constantly evolving. However, his analyses could never be adequate because he believed that one could freeze any structure and deduce its grammar ahistorically. Barthes' approach to myths largely relies on how certain icons develop their meaning within their local discourse. Second, Derrida worries that structuralism, as an objective science, is tainted by the subjective position of the structuralist. This is a common problem in scientific methodology. Yet, it never becomes a problem for Barthes since he accepts that some structures must be approached from a certain social position or subjective understanding to fully engage with them as distinct meaning apparatuses. Rather than a way of biting the bullet, Barthes is reconceiving the role of the structuralist. The examples discussed below will make this point clearer. Finally, one of the crucial points made in *Of Grammatology* is that Saussurean linguistics is phonocentric. Saussure's explicit, hierarchical organization that values speech over writing has been expressed above. For him, the signifier is a "sound-image" that only takes written form as the representation of sounds. By extending the boundaries of his semiotics, Barthes does not come into contact with the sound-image as a privileged representation of the signified. In *Elements of Semiology*, sounds are not discussed. Silence is usually endowed with more semiotic density than sound, especially in *Writing Degree Zero*. Graphic and phonic signifiers are not distinguished *prima facie* or as such.

Derrida's critique of structuralism is primarily an attack on its mode of meaning-production. *Différance*, spacing, and the trace reveal an infinite substitution that structuralism cannot account for while remaining closed. One is no longer able to say: "here are all of the terms of the structure and these are the rules by which a member may produce a well-formed utterance." Thus, since it is typically believed that meaning can only be derived when a structure is closed, Derrida is charged with abolishing all meaning in general. It is at this point that the practical applications of analysis become so vital in Barthes, who seems at almost every point to suggest the opposite of this assumption.

3

The Secret of Roland Barthes

Barthes' project varies over time and cannot be considered to be strictly structuralist at every moment. His first book, *Writing Degree Zero*, is a reaction to Jean-Paul Sartre's existential philosophy of writing. At this point, Barthes is drawing attention to variables that are important in a structural analysis, but does not become exceptionally structuralist until his later book, *Elements of Semiology*. He writes in this vein for quite a while before transitioning to a post-structuralist position. His goal in his structuralist period is to analyze structures (or what he will come to call "myths") in brief and lucid prose. However, his writings accomplish a lot more. There is an implicit project that Barthes is working at, which engenders a critique of structuralism. It is not the case that Barthes is not structuralist; he certainly is, but he is also reformulating how structuralism had been conceived by Saussure and Hjelmslev. In each of his three books discussed below, a concept is developed that pushes the boundaries of Saussurean structuralism in much the same way that Derrida attempts to do in *Of Grammatology*. While structures are put under more strain, Barthes maintains a systematic understanding of them that allows for the unequivocal production of meaning. It is this possibility that must be extracted from his works.

Langauge as Algebra, Style as Secret

Writing Degree Zero begins as a critique of writing. We have seen that Derrida is concerned about writing as well, but Barthes' discussion is focused on literary writing and its style. Barthes believed that Sartre's account of writing was incredibly reductive. According the Barthes, Sartre argues that form is not important when interpreting a text, but only the content. The structuralist reaction to this takes the form of a new approach to language. Barthes is concerned with two competing aspects that are at odds in every linguistic utterance: language and style. To follow this, one needs to be clear on a distinction made between two words used for language, which Barthes inherits from Saussure. The first, *parole*, is an individual linguistic utterance, which represents the use of language at a certain time and in a certain context. The second, *langue*, is the overarching linguistic system that contains the possibilities for *parole*; *langue* usually consists of the vocabulary and grammar of an entire language (e.g. French). So, when Barthes uses the word language, he most often means *langue*. The focus of the book is on what language means as a historically developed structure, yet there is more than just structure involved in literary creation.

Upon first glance, *Writing Degree Zero* may look like a German text in which every noun is capitalized. Barthes is not timid when it comes to this method of emphasizing a word. However, it is not pointless hyperbole. Most often, this technique calls our attention to the evolution of larger structures that can only be understood in their temporality. The term one finds most often capitalized is Literature. In his use of "Literature," Barthes is signifying a progression of linguistic possibilities that go through a process of naturalization, broken up by the occasional rupture. At any given moment, language has a finite set of well-formed utterances that have been historically determined.[1] Language is always changing through its use and the addition of new words or phrases, but at any moment there exists a finite number of possible utterances that would be understood as acceptable uses of language. According to Barthes, this gives way to a violent struggle between what the author wants to say and what is possible to say (in cases when the

1. This seems to be in direct contradiction of Noam Chomsky's argument that there are infinite possible utterances because there is no conventional limit to the length of a sentence. Chomsky is only expressing a trivial truth and has little to do with Barthes' picture of language. For Barthes, it is not the length of a sentence that is limited, but the various means of expression available to an author. It is generally accepted that some ideas cannot be expressed in a particular language. This is because each sign is caught up in a local series of differences which is necessary to derive its meaning; this is lost when one attempts to translate a sign directly. One must agree with Merleau-Ponty's point about meaning when he says that it is never translatable (cf. *Phenomenology of Perception*). Here we find one of the limitations that Barthes is directing us to in our use of language.

author seeks to say something not available in her contemporary language).[2] This struggle makes it necessary to distinguish between language and style.

Barthes aligns language with algebra. One must be careful not to take this claim metaphorically. Language, as the structuralist understands it, is a system of differences. To say something literally means to count the opposing terms and define relevant topographies. Barthes believes that there are purely algebraic uses of language, which are often found in classical literature. Classical literature is roughly defined as anything before 1850 and is most often subservient to bourgeois ideology. This is not to say that all literature before 1850 was essentially the same, but that the author's use of language (e.g. mode of expression) in those periods was extraordinarily timely. Barthes believes that literary expression in classical literature was limited to what could be said at that historical moment and rarely reached beyond language itself.[3] The appearance of avant-garde literature after the classical period (post-1850) represents a break with this historical trend.

One conclusion that could be drawn from this historiography is that the language of the author determined the ideas he expressed. Barthes finds this limitation to be characteristic of classical novels, in which bourgeois values dominated. To clarify this point, he discusses two important tropes of classical literature: the preterite and third-person singular. The preterite, which became a sort of

2. "Thus is born a tragic element in writing, since the conscious writer must henceforth fight against ancestral and all-powerful signs which, from the depths of a past foreign to him, impose Literature on him like some ritual, not like reconciliation." Roland Barthes, *Writing Degree Zero*, trans. Annette Lavers and Colin Smith (New York, NY: Hill and Wang, 1977), 86.
3. One must understand this as a failure on the part of the author to adequately express style, rather than a concrete limitation set on language.

rhetorical tool, meant to serve a certain end. He claims that it is no longer maintains the same function in spoken French and acts solely as a literary device meant to construct an intelligible Narrative.[4] The goal of the preterite is to create an artificial temporality where every event has its place as a cause and an effect, thus negating the organic spontaneity of experience. This function produces an algebra of events.[5] It instills in the reader a sense of determinism where everything in the world is as it should be; there is no possibility in a world that is always-already past. Hence, the preterite, in its complicity with bourgeois morality, encourages indifference and complacency in the face of discontent.

The third-person singular provides a similar sense of security that Barthes finds illegitimate.[6] The use of "he" or "she" provides a comparable level of coherence in the text to that of the preterite. To present a complete narrative, the author of the classical novel must externalize everything. Leaving no room for the subjectivity of the reader to enter the text, the classical novel becomes a fabric of algebraic relations. Thus, one finds a lack of meaning in classical literature. Rather than trying to say something with language, the classical novel merely reproduces the internal operations of language as such. The same conventions are repeated

---

4. "Its function is no longer that of a tense. The part it plays is to reduce reality to a point of time, and to abstract from the depth of a multiplicity of experiences, a pure verbal act, freedom from existential roots of knowledge, and directed towards a logical link with other acts, other processes, a general movement of the world: it aims at maintaining a hierarchy of facts." Barthes, *Writing Degree Zero*, 30.
5. "...it functions as the algebraic sign of an intention." ibid.
6. "The 'he' is a formal manifestation of the myth, and we have seen that, in the West at least, there is no art which does not point to its own mask. The third person, like the preterite, therefore performs this service for the art of the novel, and supplies its consumer with the security born of a credible fabrication which is yet constantly held up to be false." ibid., 35.

incessantly until meaning loses all density in an attempt to solidify the relational value of the sign within the coherence of the narrative.[7] Algebraic language's greatest fear is the "I." This fear is also where style begins to penetrate the neatly put-together structure of language.

Style is an ambiguous concept, which is no surprise since it is meant to represent an ambiguity in language. Barthes uses the word in two different ways. First, style is understood in its general sense as an individual's particular mode of speaking. Speech and writing reflect style in their minor transformations, peculiar errors, and rhythm. There is no doubt that there exists a vast difference in style between authors such as Hemingway and Faulkner. Yet one does not need to be a famous writer to have a style; one can have style by a misuse of language as well as an innovative use of it. Every appropriation of language is incomplete, since one never fully possesses the entirety of *langue*. This means that any speaker at any moment will only have knowledge of some of the rules and vocabulary of a language. This can change as the speaker learns and forgets other words or rules, or as *langue* itself changes. The way a language is acquired and the manner in which it is expressed must be different for each person. This point clarifies the second

---

7. "Overworked in a restricted number of ever-similar relations, classical words are on the way to becoming an algebra where rhetorical figures of speech, clichés, function as virtual linking devices; they have lost their density and gained a more interrelated state of speech; they operate in the manner of chemical valences, outlining a verbal area full of symmetrical connections, junctions and networks from which arise, without the respite afforded by wonder, fresh intentions towards signification. Hardly have the fragments of classical discourse yielded their meaning than they become messengers or harbingers, carrying ever further a meaning which refuses to settle within the depths of a word, but tries instead to spread widely enough to become a total gesture of intellection, that is, of communication." Barthes, *Writing Degree Zero*, 46.

way to understand style: the personal history and intentionality that exists behind any utterance. Style can be partly uncovered in how it alters the framework of language (e.g. flouting a certain grammatical principle), but it can never be completely revealed, which is why Barthes compares it to a secret.

Given its nature, Barthes attempts to explain the secret several times. According to its definition, the secret is that which is private and hidden. Language would be impossible without style.[8] Style allows language to communicate meaning and to evolve into something new so that new ideas can be expressed. Style is more than just a speaker's intention though; it also takes the form of a "secret mythology."[9] The "mythology" is the personal history of the speaker that incites language itself, but language can never access the meaning of its meaning. This would require a meta-language, which, as Barthes acknowledges in several places, would be effaced by language in its very creation.

The dimension of the secret is vertical, which also suggests a historical element. When language exists on a purely horizontal plane (as one finds in Derrida), it is the verticality of style that fixes and contextualizes utterances. The style is, therefore, an essential part of language that is necessary at all times.[10] In his praise of avant-garde literature, Barthes describes the sign liberated from all fixed connections that engages in a completely vertical expression of meaning.[11] (Of course, even this movement is subject to

---

8. Ibid., 78.
9. Ibid., 10.
10. "[A language and a style] are the natural product of Time and of the person as a biological entity." ibid., 13.
11. "Fixed connections being abolished, the word is left only with a vertical project, it is like a monolith, or a pillar which plunges into a totality of meanings, reflexes and recollections: it is a sign which stands." ibid., 47.

codification and reintegration into the algebra of language.) Style as the secret signifies Barthes' first attempt to push the boundaries of structuralism. The secret is inexpressible, but is also the foundation of language. It can be seen in short, distant glimmerings, but ultimately must disappear beneath the surface; signs are dolphins lost in the monotony of the waves, which at times decide to leap over the horizon, grinning at Poseidon's impotence.

At this juncture, a relevant distinction has been made between horizontality and verticality. Algebraic language operates on a purely horizontal plane. This point has been acknowledged and complicated by Derrida's notion of spacing. Barthes, however, is directing our attention to another facet of structures that is just as necessary. Style, in contrast to language, operates vertically on signs. We will only ever be able to read horizontally, but expression takes place vertically. When Barthes describes style as a secret mythology, one should take this to mean that the secret is ultimately decipherable. This concept will take on different names throughout Barthes' work, but the verticality of language will haunt every structure. *Elements of Semiology* is a profound analysis of the horizontality of language, but even in this work one finds Barthes defining the limits of the structuralist method only to surpass them.

The Horizontality of Language

Elements of Semiology is focused on developing a Saussurean approach to structures outside of language. It was Saussure who first proposed the term "semiology" in his *Course in General Linguistics*, which Barthes takes up in the same spirit. In *Elements*, the central approach to analysis defined by structuralism is exercised on a number of other sign-systems. By sign-system, I mean a closed structure of signification whose boundaries are determined by a sign's membership within a certain category. For example, two structures Barthes mentions are the garment industry and the food industry. It is fairly simple to distinguish an item of food as a sign in the food industry from a sign that would be common to language. However, this does not imply that ambiguities are absent from structural analysis. For the moment, we will assume that semiological structures are concrete, closed, and atemporal. Much of the book is a discussion about structuralism as it was formulated by Saussure and Hjelmslev, but Barthes is not afraid to make many new contributions to this pursuit. His additions are both corrections of his predecessors' work and adjustments of linguistic structuralism which apply it to other structures. The notion of style loses its radicalism in this book, but Barthes continues to push the boundaries of structuralism.

The formation of any sign-system depends on repetition.[12] A sign is born only through its repetition in a given milieu.[13] Each signification increases the adhesive between the signifier and signified when it is used in a particular linguistic community. Repetition, therefore, makes the meaning of a sign more clear and distinct by solidifying the bond between the signifier and signified. For ex-

12. "The combinative aspect of speech is of course of capital importance, for it implies that speech is constituted by the recurrence of identical signs: it is because signs are repeated in successive discourses and within one and the same discourse...that each sign becomes an element of the language." Roland Barthes, *Elements of Semiology*, trans. Annette Lavers and Colin Smith (New York, NY: Hill and Wang, 1977), 15.

13. Barthes uses the word "discourse," but I have found it best to use the word "structure" or agreeable synonyms to communicate this idea. After Foucault, the word "discourse" has taken on a considerably different meaning. To avoid confusion in a discussion that balances between structuralism and post-structuralism, I will avoid this word unless directly commenting on Foucault.

ample, a fashion trend gains its popularity by being repeated. To wear a baseball cap in a certain direction will signify membership in a designated sub-culture because that signifier (the direction of the cap) has been repeatedly attached to that signified (the sub-culture). This is not an extraordinary insight since Barthes is primarily reproducing Saussure's argument, but it is crucial to clarify the foundation of a closed structure if we are to understand where Derrida's assessment really gains ground. It is not long before Barthes departs from a strict adherence to Saussure.

Barthes' first departure from Saussure is on an issue that was also noteworthy for Derrida: the arbitrariness of the sign. His concern is that, in the majority of semiological systems, it is not an arbitrary mass that determines the meaning of signs, but a small group of people that decide what meanings are possible in a given language. This does not negate the possibility that signs can evolve naturally, but Barthes is also interested in the other, more dominant, aspect of semiology, which is fabricated languages or, as he calls them, "logo-techniques."[14]

The logo-technique is not forced on a speaker. Each logo-technique must be confirmed by the community through its appropriation of that sign-system. For example, if we use the example of the baseball cap again, we can understand how this fashion may be created by a corporation to sell more baseball caps (e.g. placing it in advertisements, convincing celebrities to mimic it, etc.). If the trend is appropriated, more baseball caps will be sold; if not, there is nothing the corporation can do to force this fashion on the community. Thus, through the logo-technique one finds an alternative relation between the signifier and signified, which is primarily

14. Barthes, *Elements of Semiology*, 31–2.

a motivated, contractual relation. Barthes explains that this con-
tract has two requirements: it must be collective in that it needs
to be affirmed by the majority of the community, and it must be
inscribed throughout a lengthy progression of time so that it can
take root.[15]

Likewise, Saussure believes that tradition plays an important
role in connecting a signifier to its signified. Otherwise, Saussure
would be appalled at the thought that signification could be constructed and directed by speakers, which would negate the very
relevance of langue as a structure that is independent of the speakers. For Saussure, language is organic in its arbitrariness; Barthes
counters that people will not be content with the organic alone and
will always tinker with what they can.

The next departure from Saussure is slight and more of a clarification than anything. In defining the sign, Barthes presents the
notion of a typical sign.[16] The typical sign is meant to be an um-
brella term for all types of signs, whether they are graphic, phonic,
or iconic. Here, Barthes is able to side-step Derrida's powerful argument against Saussure's phonocentrism. This move is necessary
for Barthes because some of his structural analyses do not even
include phonic signs. For example, the fashion system consists of
three realms of signs: the description of the garment, the picture
of the garment, and the real garment. There is no point in this system where anything is ever said. So, one easily finds the analytic
necessity of the typical sign. This innovation will allow Barthes to
avoid some of the traps that Saussure falls into regarding meaning-
production.

15. Ibid., 51.
16. Ibid., 47.

Barthes' account of structuralist meaning-production is unique and personal. He begins by arguing that meaning is articulation.[17] Contrary to Derrida, we should not understand this statement to suggest that "meaning is presence," but that meaning is produced. Meaning does not simply sit in the world like a rock waiting to be found; it is created through a process that involves various differential relations within a structure. At this point, it is crucial to make a distinction between signification and value, which could each pose as meaning-production, but are to be distinguished in structuralism.

Barthes adopts an example from Saussure in which a paper is cut into many dissimilar pieces. Each piece has a relation to the other pieces (value), while each piece also has its own recto and verso (signification). To make this point clear, the sign has a differential relation to all other signs within the structure (i.e. its relative value), but the sign also has an internal relation through which it becomes a sign (i.e. signifier and signified). Signification, as the articulation of the sign, engenders meaning-production. Barthes finds the paper example useful because it captures the act of simultaneously cutting out two amorphous masses from a structure.[18] This act of carving out the sign produces a cleavage between the two planes where signification is formed. The relation between the signifier and signified is not as important as the single act that produces them both within the same space. Thus the sign maintains some individual coherence despite the gap between its two terms, which is a point shared by Saussure. Barthes locates the task of semiology in this very action of cutting out. Semiology attempts to

17. Barthes, *Elements of Semiology*, 53.
18. Ibid., 56.

find the "cut here" lines on pieces of paper. The lines represent the rules of the structure, but the crucial aspect of cutting symbolizes meaning-production.

While the process of signification is clear, value has yet another valence. Value, or the way that one sign relates to another, is determined by two different kinds of relation: syntagmatic and paradigmatic. The syntagmatic value of a sign is founded in its difference from the signs that came before it. This string of signs is commonly referred to as a syntagmatic *chain*. For example, a sentence is a syntagmatic chain where signs are placed next to each other to create a value relation (e.g. an adjective modifies a noun). (Notice how the chain is represented horizontally.) Barthes makes a critical addition to this concept when he claims that syntagmatic chains do not need to be continuous and can have gaps or delays in-between signs. For example, the road-system offers street signs only when necessary and is otherwise separated by long stretches of empty road.[19] The syntagmatic is opposed to the paradigmatic, but the one is just as necessary as the other in forming a sign's value.

Graphically, the paradigmatic is usually thought to be the vertical axis since the signs that compose it are potential substitutions rather than actual signs that have been uttered. Barthes refers to the paradigmatic axis as a *field* of infinite and unordered signs that could be substituted for the current sign.[20] Whether the field of a single sign is literally infinite is hard to determine, but it would certainly be large enough to appear infinite within the time between it and the next sign. The actual operations of the chain and field

19. Ibid., 69.
20. Ibid., 81.

can get murky, so it is best to pause for an example. Consider the word "blue." It can be at the beginning or middle of a sentence, either way both relations are acting on it. The syntagmatic value of "blue" is determined by what came before and after it. Think about what word it could be modifying or what role it could possibly play in a sentence. The paradigmatic value of "blue" is determined by its possible substitutes (i.e. cerulean, red, blew). The potential substitutions that exist in the paradigm of a sign include signs that share a similar signified, signs with a similar sound-image, and signs grouped in a similar category (e.g. colors). It is clear why Barthes would claim that the potentials that exist in the field are unordered and infinite: not a single substitution seems to take precedence over another. There are virtually an infinite amount of relations a sign can hold to others given the flexibility and ambiguity of language.[21]

Barthes' understanding of value has an affinity to Derrida's *différance*. Just as *différance* differs and defers, the field does the same; the time of the trace which produces the endless echoes is mirrored by the incalculable delays of the syntagmatic chain. One must remember that *différance* characterizes a tension between the signifier and signified, so there is a slight difference here. However, the role of infinite substitution is present to the same degree. What does this mean for Barthes? Is there a finitude that can halt substitution in order to produce the meaning of an utterance? Barthes will confront this issue of time and history in another book, *Mythologies*.

---

21. The field may not be infinite in a logo-technique, which is designed with a specific purpose and limited modification of the base language. However, there is also the possibility that the logo-technique fails because the infinite nature of the field of the base language saturates the second-order signifiers of the logo-technique. Barthes, *Elements of Semiology*, 20.

The end of *Elements of Semiology* could be read as a transition
from his work on the horizontality of language to an exploration of
the verticality of history in *Mythologies*. Yet this transition does
not come without its share of tensions and contradictions. The last
paragraph embodies a tension that is not only peculiar to Barthes,
but all of structuralism. He begins by saying that, in principle, diachronic elements must be eliminated from any semiological analysis at all costs. It is better to study two objects from the same time
rather than one object over time. Nevertheless, Barthes admits that
some systems (e.g. fashion) change rapidly and the structuralist
will have to adjust to this idiosyncrasy. In this case, the structuralist
will encounter the "essential aim of semiological research," which
is the study of the time and history of a structure.[22] Thus, the
reader has gone from rejecting diachronic elements at the outset to
making them the essential aim of analysis. The problem, which is
so clearly expressed in this passage, will be resolved when Barthes
switches to an understanding of structures that relies on the concept of *myth*. History and exchange will become the paramount
conductors of meaning in *Mythologies* and most of Barthes' later
work. If *Elements of Semiology* represented an analytic investigation of algebraic language, *Mythologies* can be considered an
exploration of diachronic elements.

Myth, History, and Nature

Unlike the meandering prose of *Writing Degree Zero*, *Mythologies* has a very specific purpose. The goal of the book is to show
how the natural is derived from the historical and how the necessary is derived from the contingent. The book is divided into two

22. Ibid., 98.

parts. The first is a series of "mythologies" where Barthes takes up a certain structure and analyzes its historical origins alongside its synchronic meaning. The myth is precisely that structure which has been understood in its historical contingency. The second part of the book is an essay entitled "Myth Today" that presents a formal theory of the myth and describes the analytic methods of the mythologist. The Saussurean emphasis on synchronic elements is blind to the process that Barthes clarifies in this book, which is a gradual *naturalization* or *habituation* of signs. This process takes on a vital role in how Barthes conceives of meaning-production. To elucidate it, he brings in several of the concepts that one finds in the previous two books discussed including algebra, diachrony, Literature, and the second-order signifier. These concepts are mixed into a nuanced, cultural critique of diverse sign-systems that span from Latin grammar to wrestling. Barthes may be entering *terra incognita* with regards to structuralism, but he begins at the same place as he did in *Elements of Semiology*.

In beginning a discussion of Barthes' treatment of the myth, it is fair to understand the word in its common meaning. He is not trying to change the reader's perception of what a myth is, but merely formalize the genetic conditions and properties of a myth. Sometimes there may seem to be an unconventional use of the word "myth," but this is because its structural operation is more general than previously thought and extends beyond fairy tales or prohibitions.

The myth, like any structure, is born of repetition. In its repetition, the myth is passed through a linguistic community and gains weight in its very reproduction.[23] While it is repetition that gives life to, creates, and sustains a myth, it is also that which *naturalizes* it. Repetition heats the adhesive between signifier and signified, making it more potent. Of course, not all repetitions are successful or meaningful. It must take place within a community of speakers, in which exists the possibility for successive repetition. For example, I can use the word "cat" to denote a planet that is light-years away on which only dogs live. If I repeat the word endlessly to myself, it will not change the true denotation of "cat" once I enter the linguistic community. Language is a structure and, as such, it is always external to us and our desires. The mythologist does not study the beliefs of certain civilizations, she studies repetitions.

The first study from *Mythologies* is on wrestling. Barthes picks apart every gesture and ritual of a wrestling match to show the internal dynamics of the signs and what they produce (e.g. the pity or celebration of the spectator). The structure of the wrestling match is described as an algebra that produces its apparent causality.[24] Yet this strategy is only one way that a sign can be revealed, which is an extremely synchronic reading of a structure. The other way is to approach the history of the sign. The history of a sign

23. Two comments must be made at this juncture, one specific and one general. Saul Kripke presents a similar model of a linguistic object passing through a community of speakers, but this is very different than what Barthes has in mind. Kripke is concerned with tracing this chain of reference back to an origin; Barthes is only interested in how the myth is passed on because this (e.g. its history) is a necessary component of understanding the myth. For all our talk of chains and linguistic communities, we still refuse to invite Kripke to our dinner parties. The other point, which is more specific, is to explain what is meant by weight. The word should be understood to connote both more meaning and more complexity. This will be explained latter when considering the role of the signified in the mythic structure.
24. Barthes, *Mythologies*, 19.

is always hidden deep within itself and reveals an internal facet that is indicative of a certain moment in time.[25] Here one finds the horizontal and vertical dimensions in play again.

Either one can choose to flatten a sign and treat it as a timeless concept in an algebra of signification (c.f. *Elements of Semiology*), or one can interrogate the vertical depths of a sign, which is what Barthes sets out to do in *Mythologies*. The hidden facet of the sign, its very verticality, is history, which acts as the vehicle that delivers myths to classes of people.[26] (The myth-as-vehicle should be strictly metaphorical since a myth is closer to a virus in how it spreads.) Thus, one can understand why the transition to a mythic terminology was advantageous. A problem arises: if a structure can no longer be treated formulaically, what is its logic?

Barthes' answer to the above question is complex and rightly so. Several dynamics are at play within a single myth. First, one might ask: How is a myth internalized? Barthes nicknames this process: *Operation Margarine*. He describes it as a paradoxical procedure of discovering an evil only to accept it in the end. In this way, events become justified in their very denunciation. The internalization of a myth is not a rational discourse with an expert, but a stealth maneuver that plays on emotion and the unconscious. Barthes cites popular representations of the Army and Church as being cruel, conservative, and exhibiting the worst qualities one could associate with them. Priests are denied all human intimacy; soldiers are

25. "[Signs present themselves as] deeply rooted, invented, so to speak, on each occasion, revealing an internal, a hidden facet, and indicative of a moment in time, no longer a concept." Barthes, *Mythologies*, 28.

26. The notion of style takes an important turn in this book. It is no longer the "personal history" of a person, but the history of a class or group of people. People can embody sign-systems, but those systems are communal and determined externally.

turned into indifferent killing machines. Despite the surface signification of any such representation, the viewer is able sympathize
with and recognize a transcendent power in these institutions despite their evil side. For example, *Full Metal Jacket* reveals the
detestable conditions of military training and combat during the
Vietnam War, but by the end of the movie the spectator finds
those evils necessary to produce the camaraderie of the soldiers.
The last scene offers complete absolution to the U.S. government,
drill sergeants, and other culpable parties; the soldiers were able
to maintain their innocence as they march through charred fields
singing the Mickey Mouse jingle. It is not that misdeeds have been
forgotten, but that they are re-interpreted. A myth is a reposition-
ing and a distancing: the signified takes a step back and becomes
fuzzy.[27] As such, the myth relies on time to become what it is.

TThe crucial quality of any myth is that it is historically contingent. Like language, or any other structure, it evolves over time.
The mythologist is a sensitive lover, searching for every scar and
birthmark on the body of the myth. To begin explaining the historical foundations of the myth, it would be good to look at Barthes'
analysis of the writing of history. In the chapter titled "The *Blue
Guide*," Barthes approaches the common French guide book and
notices that the extent of its cultural knowledge is found in numerous pictures of monuments. With this observation, Barthes declares
that the guide book does not speak to anything in the present since

---

27. This notion shares many valences with Foucault's discourse formation.
The myth develops like a gloss, an extra layering of signifiers that re-
contextualize an event. For example, the discourse of sexuality is set in the
religious-psychoanalytic form of confession. Yet, we must not become apa-
thetic to the seriousness of this argument simply because the examples are not
our own. America is ripe with myths: one does not see pictures of dead Iraqi
children without immediately seeing complicity with terrorists.

it suppresses the reality of the people inhabiting that region by displaying only monuments. This gesture, however, also suppresses
the historical since the present is always already historical.[28] The
strategic arrangement of objects in the guide book, as Barthes argues, suggests an eighteenth century bourgeois mythology where
a certain geography of value was adopted based on the new found
freedom in buying. Thus, every sight becomes a sign emblematic
of and inextricable from this history. For example, the mention of
a tunnel is sufficient to represent a mountain along a scenic route.

The purpose of the chapter on the *Blue Guide* is to show that
the sign always says more than one expects it to say. All signs
carry a history with them, but this should not be understood in the
form of a symbol.[29] The history of a sign pushes in on the sign.
In a myth, it takes the form of a second-order signifier that sits
like a gloss over the former sign. However, it is not a layering-on
of signifiers that gives the sign its weight. The signified, which
is a functional selection within a given structure, determines the
role the sign will play. Signifiers are everywhere, but it is only
when they become attached to a signified that they can produce
meaning. Barthes uses the example of roses.[30] The roses acquire a
signified within a context (e.g. Valentine's Day). However, the rose
has an entirely different weight at its origin. The field hand does

28. "To select only monuments suppresses at one stroke the reality of the
land and that of its people, it accounts for nothing of the present, that is,
nothing historical, and as a consequence, the monuments themselves become
undecipherable, therefore senseless." Barthes, *Mythologies*, 76.
29. The symbol *represents* another object from which is derived its meaning.
The engraving of Napoleon that Julien Sorel covets in *The Red and the Black*
is a symbol of his ability to earn a fortune despite his humble origins. The sign,
on the other hand, has no meaning until it enters a system of differences.
30. I will depart slightly from Barthes' use of the example. Barthes, *Mythologies*, 113.

not approach the rose as an object of passion, but as a dangerous obstacle that demands care and precision. The way a signifier is attached to a signified is always through a historical situation, which also explains its variance.

The myth can never be reduced to some sort of natural phenomenon and is always constructed through an internal process.[31] Myth necessarily distorts nature by being grafted onto it and suggesting new valences of signifiers.[32] It becomes a signifying apparatus in itself, in which it illuminates the very complicated relationship between the signifier and signified. Furthermore, mythology displays a continuation of Barthes' argument regarding logotechniques (i.e. motivated semiotic systems). In "Myth Today," Barthes focuses on two illuminating myths that reveal their motivations to the mythologist.

The first example Barthes refers to is an observation he takes from Paul Valéry. A high school student opens his Latin grammar text book and finds the line: *quia ego nominor leo* (because my name is lion). It is a simple sentence, but is meant to show something more that is gradually revealed to the student. It is a lesson on how the predicates (*ego* and *leo*) must agree. One could imagine a similar English example, this time as an error that needs correcting: "I have five staple." The word "staple" does not agree with the quantifier and presents a problem that the student will learn from. Barthes explains how this reveals the existence of a greater semiological system dormant in the grammar example. We are being told something that is not present on the surface, but is derived from context and history.

31. Ibid., 110.
32. Barthes refers to this process as "language-robbery." ibid., 131.

The other example is the cover of a magazine, *Paris-Match*, which shows a "young Negro" in French uniform saluting the French flag. This scene is indicative of the French political climate at the time, so its meaning may be lost. An equivalent American example would be a Muslim, dressed as a marine, saluting the American flag. In both instances, a person who is perceived as an outsider to the country is actively supporting that country. Despite whatever political calamities (e.g. imperialism, colonialism) may be terrorizing the foreign country that this young person came from, he still honors French-ness in a very militaristic fashion. On the surface is nationalism; beneath the veneer is racism. Barthes reads the subtext: the foreigner is transformed from a savage to a citizen. As such, the foreigner now relies on the French government for stability and, most egregiously, he pledges his life for France. Every master believes his slave goes to bed at night thankful for the master's generosity. In similar fashion, this illusion manifests itself on the cover of a popular magazine to serve as an *apologia* for exploitation, an *apologia* that is neither said nor heard. Such is the terrifying nature of myth—to not see what something really is. It is naïve to think that myth is merely a currency that is exchanged and not something that one embodies. This kaleidoscopic dispersion will characterize the production of meaning within the myth.

The myth produces meaning in a fundamentally different way than other semiotic systems. Meaning-production took place on two planes in *Elements of Semiology*. Yet when one considers the effect of history on the sign, each moment can potentially multiply the number of planes. Insofar as the sign carries with it the history of its repetitions (i.e. its weight), it also holds within itself the multiplicity of paradigmatic relations that it had previously embodied.

So, meaning becomes more than mere articulation in the moment. It is rather the case that it prompts an entire system of values, current and historical, to arise in the comprehension of sign.[33]

In addition to the myth's internal history, it enters a relation of play between itself and the social groups that interact with it. To appropriate a myth, one must already be a member of a certain group.[34] Recall the two examples from the previous paragraph: the entailed classes were Latin grammar students and French citizens. A game of hide-and-seek between signifying apparatuses is how the myth comes to be. The myth exists as an objective structure, but can only be understood from a particular subject-position within time. Therefore, the myth is categorically distinct from other structures because its meaning-production takes place vertically: either through a historical alliance of exterior signifying apparatuses or with respect to the engaged subject. In this way, *meaning is contingent*. The mode of meaning-production is further complicated by its dependence on the two external conditions. Nevertheless, we must continue to understand meaning in its structuralist sense. Perhaps a more adequate word in this sense would be *significance*, but a shift in vocabulary will not alleviate the force of this conclusion.[35] This new understanding of meaning-production will guide Barthes towards a different kind of structuralism.

In the chapter "The Nautilus and the Drunken Boat," Barthes contrasts Jules Verne to Arthur Rimbaud. While Verne remained obsessed with cataloging and accounting for what there is, Rim-

33. "The meaning contained a whole system of values: a history, a geography, a morality, a zoology, a Literature." Barthes, *Mythologies*, 118.
34. Ibid., 119.
35. Barthes claims there is no meaning, only significance. Barthes, *Writing Degree Zero*, 1.

baud sought to derange his senses with what was unknown and disturbing. Here one finds two faces of structuralism: closed and open. Closed structuralism is obsessed with horizontality (e.g. the *Nautilus*, Derrida's spacing, and algebra) and deriving value from contemporary differences in a definite field of terms. Open structuralism (or what I will later call "hyphology"), on the other hand, recognizes unimaginable, vertical heights that are essential in any structure (e.g. history, secret, and context). Closed structuralism demands that the structure remain on a plateau; open structuralism will sometimes plateau the structure for convenience, but will continue to explore its "differential geology" while carving out what meanings it finds useful.[36]

Through all of Barthes' works that I have discussed so far, the boundaries of structuralism are expanded and stretched. He begins to hint at a new structuralist methodology, but is never able to develop it and, instead, loses himself in post-structuralism. For both Derrida and Barthes, repetition serves as the foundation for structuralism. This repetition is conceived in a completely horizontal manner. Hyphology, if it is to survive deconstruction, requires a vertical conception of repetition that could accommodate the production of meaning without getting lost in infinite substitution. The reconceptualization of repetition begins with Kierkegaard and comes to fruition in Gilles Deleuze, but it is only through phenomenology that a reconsideration of structural repetition can occur. It is now to a foreign land that we venture in hopes of saving our own.

36. Barthes, *Mythologies*, 76.

4

The Problem of Repetition

It is possible that this entire time we have been discussing repetition. One finds that repetition is the sense-making mechanism of signs, structures, and myths. It is paramount in Saussure, Barthes, and Derrida through the concepts of signification, structuring, and trace, respectively. I signify because I repeat. However, it is clear that a certain understanding of repetition is being expressed here. This particular position interprets the role of repetition as the reoccurrence of the same. Sign x is uttered at time t_0 and then again at t_1: this constitutes a repetition of sign x. So, repetition in this sense serves two purposes. First, it is to solidify the identity of an object (e.g. sign). Sign x becomes more clear and distinct because it has appeared in different contexts or at different times; one is better able to determine the borders of the sign. Second, repetition connects two events by a rule of association. If the object at t_1 is sufficiently similar to that at t_0, then it could be claimed that the object has been repeated. (These revelations on repetition should not surprise anyone, but are necessary in setting up the arguments that will follow.) At this point, one must ask, along with Kierkegaard: what happens to the repeated object?[1] Does it remain the same or does it change slightly with each repetition? Although repetition implies the reoccurrence of the same, it seems fair to ask if the passage of time or re-apprehension of the object can change it at all. Barthes attempts to answer this question in two different ways. It is because of the vital importance of repetition in structuralism that Barthes' explicit comments on this topic will be revealing.

Repetition in Barthes' Middle Works

The first way that Barthes answers the question of the repeated object is entirely negative. It would appear that this response tries to be as true to structuralism as possible, which can only be clear once his second response is explicated. The first response is structurally and ethically negative. Barthes dedicates a lengthy section

---

1. Kierkegaard's *Repetition* was a revolutionary book in the development of a new conception of repetition. He begins with the problem of what a repetition, which is considered objectively and subjectively, means. These are the questions that I wish to begin with now that we have clarified the traditional understanding of repetition. While this essay will not include a detailed examination of Kierkegaard's view, it is important to keep in mind that this is where the project begins.

of *The Pleasure of the Text* to explaining these negative modes of repetition. It is difficult to sort out how this discussion relates to the role repetition plays in the early works, but one must keep in mind that Barthes begins to entertain the notion of escaping structuralism. This is not to say that his critique of repetition is post-structuralist, but that Barthes is once again pushing the boundaries of traditional structuralism.

Structurally, repetition is understood as stymieing structures. The natural evolution of a structure is stunted by the asymptotic repetition which tends toward ideology.[2] Barthes believes this process to be connected to meaning. Repetition, as we saw in the case of mythologies, makes people feel that contingent, historical meanings are actually necessary, natural truths. The fact of the matter is negated in its mythology (e.g. the "Negro" saluting the French flag). So, repetition leads to the coagulation of a structure in which new meanings are aborted. This movement constitutes the logic of naturalization that is seen in mythology.

Ethically, Barthes opposes repetition because he sees it as being complicit with bourgeois ideology in the form of "stereotypes." Culture is the product of a *mass banalization* of everyday life through constructed identities and repetitive forms that underlie distractions. There are at any one time a finite number of stereotypes. A casual stroll through the mall shows that individuality is the greatest myth of all. The production of goods is constantly funneled

2. "Now, encratic language (the language produced and spread under the protection of power) is statutorily a language of repetition; all official institutions of language are repeating machines: school, sports, advertising, popular songs, news, all continually repeat the same structure, the same meaning, often the same words: the stereotype is a political fact, the major figure of ideology." Roland Barthes, *The Pleasure of the Text*, trans. Richard Miller (New York, NY: Hill and Wang, 1975), 40.

into set identities that the consumer is forced to inhabit: one must literally choose between playing a social role and living naked in the jungle.

Repetition is what produces stereotypes in how it reiterates a configuration of signs until it becomes naturalized. For example, action movies are endlessly reproduced with the same form to assure that a certain stereotype (e.g. the action movie viewer) is satisfied. Stereotypes necessarily lead to an implicit pleasure in repetition.[3] The new becomes frightening because it represents itself as the outside or the Other in a structure that should be closed in on itself; repetition is complicit with this closure. To explain this, let's look at the action movie example again.

The action movie stereotype represents a certain configuration of signs that gives pleasure to its viewer in each of its manifestations. An experimental film would exist outside of this repeated form and would require the viewer to appropriate or negate it from her entrenched stereotype.[4] This is precisely Barthes' issue with repetition: it is responsible for closing a structure (via coagulation, naturalization, mythologizing, etc.), which necessarily negates anything outside of that structure. Ethically and structurally, Barthes is beginning to realize that repetition obscures its object by isolating it. The object is isolated because it is enveloped within a closure; it is obscured because repetition distorts its organic existence within a heterogeneous environment. Barthes' affirmation of the New against solid structures produced by repetition is similar to his critique of ahistorical myths; repetition merely represents

3. Barthes, *Pleasure of the Text*, 42.
4. "The petit-bourgeois is a man unable to imagine the Other. If he comes face to face with him, he blinds himself, ignores and denies him, or else transforms him into himself." Barthes, *Mythologies*, 151.

another process of naturalization. However, Barthes begins to re-
conceptualize the role of repetition in S/Z where, rather than closing structures, repetition is responsible for opening new possibilities
of signification.

S/Z is essentially a book of two parts. First, Barthes quickly
outlines a new way of reading (and re-reading) a text. In this book,
the text that Barthes decides to analyze is a short story called "Sarrasine" by Honoré de Balzac. As Barthes clarifies the conditions by
which he will analyze the story, he is also reformulating meaning-
production as it is determined by structuralism.[5] The second part
of the book is an analytic engagement with the short story that
analyzes it line by line. Each line is interpreted with one or more of
the five codes that Barthes explicates in the first part of the book.
For example, one code organizes cultural significations, which includes the scientific or common knowledge of a culture at a given
historical time. The interpretation of the text through the codes is
meant to produce a consistent and meaningful reading of the text
by accounting for its myriad of interwoven signifying chains. However, this interpretation is only one reading of the text where there
are several other equally meaningful readings available. To explain
this multiplicity of interpretations, Barthes describes a process of
re-reading that is crucial to meaning-production.

To emphasize the role of re-reading, Barthes argues that the

---

5. As one finds in the previous work discussed, Barthes is making a transition
in S/Z from structuralism to post-structuralism. I have tried to avoid the term
"post-structuralism" in the main text of this essay to prevent confusion about
what is being argued. Our aim has never been to show how Barthes grows
into his post-structuralist phase, but to elaborate the progress he makes in
structuralism. What I will eventually call "hyphology" is not a form of post-
structuralism, but is somewhere in-between early Barthes and Derrida.

text is never whole or absolute.[6] The text is a plurality of signifiers that must be distinguished from its exterior and its totality. Each reading is not a structure in itself, but a structuration.[7] What this means is that each reading is a decision to organize the signs of the text in a certain way and to emphasize certain codes over others within specific passages. Thus, the reader is responsible for the meaning of the text.[8] The operation of re-reading and organizing each signifier but a different code multiplies the quantity of signifiers in the text. The mailbox labeled "W.A.S.T.E." will become a new signifier with each unique reading.[9] This process necessarily obscures any truth or absolute interpretation of the text because one cannot qualitatively distinguish between signifiers.[10]

Barthes argues that re-reading is an approach to the text on its own level, that is, as a plurality. He aligns the stigma against re-reading with bourgeois, consumer culture because the logic of capital wants you to buy more books rather than re-read old ones; re-reading is thought to be radical in this sense. It opens up the reader to the play of the text, which is not only a play between interpretations, but also signifiers, signs, codes, metonymies, and folds. Of course, meaning-production is affected by a structuration without nature or origin that comes to be through chaotic forces.

6. "...nothing exists outside the text, there is never a whole of the text..." Roland Barthes, *S/Z: An Essay*, trans. Richard Miller (New York, NY: Hill and Wang, 1974), 6.
7. Ibid., 20.
8. Ibid., 4.
9. The object used in this example is completely irrelevant, but it has been adopted from Thomas Pynchon's The Crying of Lot 49.
10. "Thus it would be wrong to say that if we undertake to reread the text we do so for some intellectual advantage (to understand better, to analyze on good grounds): it is actually and invariably for a ludic advantage: to multiply the signifiers, not to reach some ultimate signified." Barthes, *S/Z*, 62, 165.

Barthes likens meaning to an inflexible barrier, but also to a tran-scribability that reflects the infinite and circular relations within the plurality of systems.[11] Yet this plurality can never be closed, which would end up producing a "theological meaning" of the text.[12]

The reader must continually forget old readings to be able to read again. Signification would be impossible if one ascribed mul-tiple codes to a signifier since it would not be able to designate a single or proper signified. It can be argued that contesting in-terpretations can be kept in mind while approaching a text, but these interpretations do not inhabit the text as the current reading of it does. Take, for example, the final line of "The Swimmer" by John Cheever: "He shouted, pounded on the door, tried to force it with his shoulder, and then, looking in at the windows, saw that the place was empty." I can entertain possible interpretations of this line that are different than my own, but these interpretations are other texts that relate to wholly different signifiers. One must remember that each reading re-inscribes the word into new config-urations of codes, which makes them completely new signifiers in a unique system of differences. The empty house of the swimmer is a signifier that is altered by the code it inhabits, which also depends on the chain of embedded signifiers that came before.

If it is beginning to look like there could be an infinite amount of legitimate interpretations of any given text because of the dense plurality of variables, this is precisely what Barthes is trying to show us. There is virtually no end to re-reading, no mastery of the text that is ever an absolute exhaustion of it. One can feel done with interpreting a story, but this never is indicative of an

11. Ibid., 65, 120.
12. Ibid., 11.

end to interpretation. It is this concept of re-reading that is clos-
est to a re-conceptualization of repetition. Barthes is not afraid
to even call repetition the "fact that there is no reason to stop."[13]
With Barthes' comments in mind, we will now turn to the phenomenological precursors that pave the way for a new structuralist
conception of repetition.

The Genesis of Repetition: Merleau-Ponty and Simondon

To enter into a more profound understanding of repetition, one
must first begin to understand how this notion of repetition is even
accessible. This other kind of repetition (which will acquire a name
of its own soon) spreads itself across time like a cat stretching out.
Time and history, as Barthes had illuminated them, engender a
vertical dimension of structures. In this verticality (that sometimes
goes by the names *secret* or *field*), a different kind of repetition is
found. Naturally, it is a structural process, but one is only able to
encounter this process through phenomenology. Only phenomenology provides a bridge to memory, which is vital when interpreting
temporal processes like repetition. The phenomenological aspect
will also be germane to our discussion of Deleuze, who transitions
between the structure and its phenomenological genesis. So, it
should be clear to the reader that the goal to re-interpret repetition does not involve simply inserting a phenomenological understanding of that concept into structuralism. This would be beyond
perverse and ultimately fallacious. What we purpose is a detour
through the senses to find the virtual structures which compose
the actual with the intent of drawing out the vertical movement of

13. Barthes, *S/Z*, 177.

repetition. There are several entry points to this discussion. I will
begin where quite a few of them are gathered.

Maurice Merleau-Ponty's phenomenology will be important at
this point, but it is also interesting to note his position in the larger
development of philosophical thought. He would be heavily influenced by structuralism and purpose a structural theory of language
(cf. *The Prose of the World*). However, one finds concepts in
his phenomenology that establish a world that is ripe for structural
analysis. In *Phenomenology of Perception*, Merleau-Ponty creates
the term "phenomenal field" to describe one's perception of the
world at a given time. What is interesting about this concept is
that the phenomenal field exists as an organism of its own. Each
object in the world is a product of the internal coherence of its
forces. It is neither the rationalist subject nor the empiricist object that is responsible for the act of perception; there is not one
dominating force that captures the other, but two organisms that
co-exist in a world full of organisms. For example, a plant *appears to me* as having certain properties, but is much more than
its exterior qualities. The color of a plant tells me nothing about
the process of photosynthesis. On the other hand, the plant can
respond to me, such as when I water it, but it lacks the ability
to understand my digestive system. In this way, there are always
systems behind (subjective) perceptions of objects that help man-
ifest the qualities of those perceptions. Merleau-Ponty compared
these hidden, systemic operations to "folds," which can partly be
revealed at times, but never entirely. If an absolute perception of
an object exists, it can never be discovered since it is masked by
horizons of perception. These limits of perception do not block the
natural world from being explored, but expose the very possibility

of exploration.

The phenomenal field already establishes an object for structural analysis, but there is one other important concept that Merleau-Ponty develops. The figure-ground relation is not a new concept in phenomenology when Merleau-Ponty writes about it, but he does add to it. This relation posits that a ground (as in background) is required to distinguish an object from other objects. For example, I can tell the apple apart from the tree it grows on because the apple is a figure distinguished from the tree-as-ground. What Merleau-Ponty says about this relation is that it involves more than just the two terms. The figure and ground have an intimate relation that is more involved than a simple contrast. The ground indirectly affects the perception of the figure. So, a background effect greatly influences how a figure is perceived. It is possible to see very different objects with only a change in, say, the light's intensity. To explain this relation and how it becomes transformed by later philosophers, I will use a common example: lightning. For Merleau-Ponty, the lightning is the figure that hovers above the sky-as-ground. The lightning takes on its super-bright quality not only because it is electricity, but also because it is set in relation to the dark sky. The halo of light around the storm is not inherent in lightning itself, but is a product of its relation to a certain ground. The ground as an internally coherent phenomenal field is where Merleau-Ponty's student, Gilbert Simondon, will lead this line of thought towards structuralism.

It is not surprising that Simondon, as Merleau-Ponty's student, adopts a lot of the same positions. Within Simondon, one immediately finds conceptual equivalents to the phenomenal field, internal coherence, the organic nature of the object, the figure-ground relation, and also passionate rejections of rationalism and empiricism.[14] He is interested in finding the structural conditions for physical and biological genesis. Simondon realizes, like Merleau-Ponty, that a phenomenon is only a single appearance of an object that is more complex. He will use the concept of *individuation* to describe the process whereby an object becomes what it is as it appears to us. Therefore, where Merleau-Ponty is only interested in the phenomenology of perception, Simondon takes up the task of investigating the structural potentiality of the individuated object.

Simondon moves beyond Merleau-Ponty's phenomenology in two important ways. The first is his understanding of the subject. Simondon agrees that the subject cannot be comprehended in isolation and is necessarily part of an environment in which it has a symbiotic relationship with objects.[15] However, this conception departs from the phenomenological when it disavows Merleau-Ponty's emphasis on ambiguity. Simondon believes that the encounter with an object represents the individuation of a field of intensities that is definite in nature. The object represents a reality of intensities breaching various thresholds.[16] So, the phenomenal field is no longer a desert of ambiguity, but a chaotic graph in which each

14. It is relevant to note that Simondon is only recently gaining recognition in the English-speaking world. This means that most translations of his work (what few there are) have been done in short form by independent scholars. The text I refer to exclusively is the first chapter of Simondon's *Psychic and Collective Individuation*, which was translated by Taylor Adkins and posted to his blog "Speculative Heresy."

15. "The entire subject should be considered in a concrete situation with tendencies, instincts, and passions, and not the subject in a laboratory, in a situation that has a weak emotive valorization in general." Gilbert Simondon, "Translation: Chapter 1 of Simondon's *Psychic and Collective Individuation*," Speculative Heresy, §5, accessed October 6, 2008, http://speculativehere sy.wordpress.com/2008/10/06/translation-chapter-1-of-simondons-psychic-and-collective-individuation/.

16. Ibid., §4.

point possesses a spatio-temporal location. The subject's role is also magnified by Simondon. He argues that it is the act of perception which organizes the object into a coherent whole in relation to the subject; each perception is an orientation to the world.[17] Simondon agrees with Merleau-Ponty that perception involves organizing the world and making sense of the nature of objects. However, Simondon views this function as the most primordial activity of the subject.

The second point that Simondon moves away from his teacher on is the notion of *pregnancy*. Some of this has already been hinted at already. For Merleau-Ponty, the object is pregnant with meaning because it always holds more: more folds, more ambiguity, more perspectives, and more horizons. Simondon replaces this notion with that of *intensity*. Intensity refers to the information received by the subject concerning the quantities and qualities of the object. Intensities can only be distinguished according to relative differences. For instance, I can say that a plant needs water because I witness a decrease in a specific intensity (i.e. its green color). An intensity in itself, abstracted from its phenomenal field and history, would make no sense.

Each object arises from a space of possibility in which can be found its conditions for being. The appearance of an object is determined by its level of intensity. Intensities are distinguished by thresholds that produce a change (e.g. chemical phase-shift) in the object. The intensity picture that Simondon develops is more believable because it explains how objects change, decay, and grow. In Merleau-Ponty, objects are ambiguous and infinite, but they are

17. "The subject perceives so as to be oriented in relation to the world." Simondon, *"Psychic and Collective Individuation,"* §5.

all-too-inert and reliant on the subject. The internal coherence of the plant, as discussed above, can only explain the appearance of the plant as it is *now*. As far as the phenomenology of perception is concerned, we learn nothing about what the plant can do or what it will become.

Simondon sets out to explain how objects change over time, such as the complex phenomenon of crystallization in a super-saturated solution. The form of the crystals are not inherent in the homogenous, liquid solution, but are produced out of it. Simondon claims that this process is indicative of the functioning of perception; like the super-saturated solution, the world is meta-stable.[18] In this sense, Simondon makes the crucial move toward understanding the diachronic nature of the object.[19]

Intensities are able to inform a subject about more than just the contemporary state of an object. They also designate the potentiality of an object in that the object is pushed toward a certain probabilistic future. For example, the sex of a reptile egg is determined by the temperature of the environment during the period of gestation. The temperature is a force that alters imperceptible intensities within the embryo. If the temperature pushes those intensities over a certain threshold, the egg will become a specific sex. To be clear, it is not *more* or *stronger* intensities that the environment (e.g. temperature) determines, but the *density* of already resonating intensities within the object (e.g. embryo).

18. "there is also an intensive manifold that renders the subject-world system comparable to a supersaturated solution; perception is the resolution that transforms the tensions that affected this supersaturated system into an organized structure; it could be said that every true perception is the resolution of a problem of compatibility." ibid.
19. Merleau-Ponty is concerned with temporality as well, but only as it reflects on the subject.

Hence, Simondon is able to articulate processes that seemed obscure or irrelevant in the past. Nevertheless, he will have to posit a distinction between the space of intensities and that of objects in order to explain how individuation occurs.

Simondon appropriates a distinction made by Henri Bergson between the actual and the virtual. This distinction will be incredibly important for both Simondon and Deleuze. As Bergson had argued, the actual and virtual compose the totality of the Real. The virtual is just as real as the actual, but operates on a different plane of reality. Whereas the actual is what appears, the virtual is what produces that appearance. This means that there is something behind the appearance that makes it what it is. For Simondon, the virtual is inhabited by intensities and thresholds. So, one should not immediately reduce this distinction to the real and the possible. Intensities exist concurrently with their individuated objects; possibilities do not exist concurrently. Like Saussure's paper analogy for the signifier/signified relation, actuality and virtuality relate to each other like the two faces of a single sheet of paper.[20]

The process through which the virtual becomes actual is called *actualization*. Actualization engenders a play between the actual and virtual whereby they influence each other. This is clearly seen in the example of the reptile egg where the temperature affects the intensities, which then determine the gender. Here the actual temperature affects virtual intensities that then determine the actual sex. In a sort of dialogic relation, tensions within the virtual are responded to by the actual and vice versa. This process creates

---

20. This analogy is helpful for understanding how the virtual can be a different plane of reality, but still exist in the real. However, one should not make too much out of this comparison. Actuality and virtuality simply cannot be reduced to the signifier-signified relation.

an inherent meaning in the progression of reality. The actual responds to the virtual like a solution relates to a problem. Changes
in actuality are solutions, but also potentially create new problems
as a new configuration of the phenomenal field develops. However,
each configuration holds meaning in that it presents an answer to a
problem.[21] The play of actualization and meaning-production will
become especially important as we turn toward Deleuze.

The similarities between Simondon and Deleuze are numerous
and we have not yet exhausted them. Simondon is a perfect middle-
ground between Merleau-Ponty and Deleuze. He is able to carry
the emphasis on meaning and temporality that one finds in phenomenology to structuralism, which is the main task of this critique
of Derrida. Simondon is an Egyptologist, bringing us to hieroglyphs
never seen before, but one requires Deleuze to decipher them. Simondon has phenomenologically hinted at a structuralism of the
virtual, but this project is only fully comprehended and completed
in Deleuze.

Deleuze's Secret Repetition

This chapter has been leading up to the re-conceptualization of
repetition that Deleuze makes in *Difference and Repetition*. He inherits much of its rethinking of the concept from previous theorists
such as Kierkegaard, Nietzsche, and Gabriel Tarde. While others
(e.g. Heidegger and Freud) have been influenced by this challenge
to the traditional conception of repetition, Deleuze takes it in a
unique direction. He attempts to think of repetition as a double

21. "The object is an exceptional reality; in a quotidian way, it is not the object
that is perceived, but the world, polarized in such a way that the situation has
a meaning." Simondon, "*Psychic and Collective Individuation*," §5.

movement. This movement takes place in structures and perception and is known as masked and bare repetition, respectively. This dual aspect of repetition is what I intend to clarify. By reinventing our understanding of repetition, we can approach structuralism in a non-limiting manner.

The development of Deleuze's project relies a lot on the precursors outlined above, especially Simondon. It even bases itself in the same Bergsonian distinction between the virtual and the actual. Deleuze's intent, however, is much different. His goal is to uncover the structure of the virtual insofar as it manifests the conditions for signification. This may sound fairly structuralist, and it should. Deleuze is ultimately doing a kind of structuralism, but through different methods. He is attempting to access the virtual through a phenomenology of the actual. He realizes that the virtual, as Simondon had shown, is a pulsating mass of intensities that produces individuated objects. Yet he is also interested in how the process of actualization sublimates the differential forces in the virtual to create associations within the actual such as representation, identity, and similarity.[22] In its demonstration of how identity (actuality) derives from difference (virtuality), *Difference and Repetition* leads us into the heart of darkness that is behind the perceivable world.

The story begins in Chapter Two, which is entitled "Repetition

22. Deleuze's focus on representation is similar to what Derrida was doing in *Of Grammatology*. However, their projects have different ends. Derrida wanted to show that representation was problematic because the represented object could never be present. Representation gave way to an endless play of signifiers. For Deleuze, representation is not internally problematic, but is a way of organizing the actual and a by-product of individuation. So, there are two points being made: (1) representation (in the form of representational realism) is a contingent philosophical position, and (2) representation is treated as an associative relation between individuated objects that does not account for their becoming in the virtual.

for Itself." Here one finds an explanation of how one gets from the sensible apprehension of matter to thought. The nature of this *Bildung* is not entirely settled between scholars of Deleuze. It is up for debate whether the story takes place in the realm of ontology or whether it begins with the phenomenological positing of an originary subject.[23] Regardless of this issue, the current task is simply to understand how Deleuze transforms the notion of repetition.

The transformation begins with a simple distinction between active and passive syntheses. These syntheses unfold in time and constituted our lived experience. Active syntheses involve acts of consciousness that express intention. For instance, focusing on a certain object or remembering a specific event. Passive syntheses are opposed to active because they take place beneath conscious perception, but contribute to what is perceived. While a passive synthesis is not an action done by a subject, it still "occurs *in* the mind, which contemplates, prior to all memory and reflection."[24] Deleuze outlines three passive syntheses of time that are responsible for constituting the living present.

The first passive synthesis begins with sensibility. My immediate sensation of the world constitutes a moment in time. This moment is defined as none other than the present. Contrary to the Aristotelian notion that time is a successive series of nows, Deleuze suggests that our present is limited by fatigue. Only so many moments of immediate sensation can be held in contemplation before

---

23. For an ontological interpretation, see Daniel Smith and John Protevi, "Gilles Deleuze," The Stanford Encyclopedia of Philosophy, accessed January 14, 2020, http://plato.stanford.edu/archives/spr2013/entries/deleuze/; for a phenomenological interpretation, see Joe Hughes, *Deleuze and the Genesis of Representation* (London: Continuum, 2008), 132.

24. Gilles Deleuze, *Difference and Repetition*, trans. Paul Patton (New York, NY: Columbia University Press, 1994), 71.

the mind tires and forgets. As soon as it forgets, however, it picks up the thread again and begins to hold together another series of sensations. Each set of sensations is called a *contraction*.

Contractions, as the synthesis of immediate sensations, are what make up the present. One must be careful to notice the plurality of sensations that each contraction can hold. Bergson contends that present moments are made up of singularities as if each now were a tick on the clock. Deleuze represents this understanding of the series as the repetition of a singular event: A, A, A. In contrast to Bergson, Deleuze favors a supposedly Humean representation of contractions in the form of AB, AB, AB. In this case, a contraction holds within it differential elements (A and B) rather than a singular, homogenous moment.

The nature of the contraction is essential to understanding Deleuze's distinction between two kinds of repetition. For repetition to occur, a case must already be established. The "case" for Deleuze is the contraction, which is a set of differential elements in the form of various immediate sensations. The elements must be arranged in the same way during a secondary occurrence in order for repetition to occur. Hence, AB_1 is a repetition of AB_0 if and only if the same differential elements (A then B) occur within the same contraction. Deleuze will sometimes say that this passive synthesis constitutes habit and imagination. One can see that the repetition of AB would induce a habit to predict that a B will follow every A, in the same way that smoke follows fire. The habit of anticipating B after A is something that occurs, according to Hume, in the imagination. However, this has only described one kind of repetition. Deleuze's brilliant move comes in designating the other, more secretive kind of repetition.

The repetition described above is sometimes referred to by Deleuze as the "repetition of the same" or "bare repetition." Its object is the elements that exist within a contraction. This form of repetition would not be possible, however, without a more fundamental and structural repetition. This other repetition is the repetition of contractions themselves. While bare repetition merely acknowledges the reproduction of the same representation, masked repetition "never ceases to unravel itself." There may never be another contraction of the form AB, but there will always be another contraction. This necessity is constituted by a structural repetition.

Bare repetition is like "a skin which unravels" to expose the complex, internal repetitions of contractions.[25] Deleuze uses the example of a chicken pecking at grain on the ground to explain this relation. There is a passive organic synthesis, he argues, that connects the chicken's cardiac pulsations with its head nods. This passive synthesis forms the ground for the active, perceptual synthesis that allows the chicken to peck at grain. This example is illuminating for two reasons. First, it reminds us that the first passive synthesis of time is unconscious. While I have mentioned habit and imagination, these are allusions to the actual, active processes they usually denote. The first passive synthesis forms the ground from which these faculties can emerge. Second, a relation of difference exists between the two repetitive syntheses in the chicken. It is only through the internal differentiation of the two repetitions that anything like experience can arise. It is for this reason that Deleuze claims, "Difference inhabits repetition." In other words, it is only from this fundamental difference that anything like "repetition of the same" can occur. Already, we are getting a glimpse

25. Deleuze, *Difference and Repetition*, 76.

of how masked repetition is meant to challenge the privileging of
identity in traditional repetition.

The first passive synthesis of time was responsible for creating
the living present. Yet it is not possible for the story to end here.
The living present, as we have seen, is always already a contraction limited by fatigue.[26] As soon as it passes away, so does that
present instant. Taking up a new contraction is akin to inhabiting
a new present. One must ask: Where do the old contractions go?
This question is answered by the second passive synthesis, which
establishes the past.

In the same way that the first passive synthesis was broken up
into two series (cases/elements), the second passive synthesis articulates two forms of the past. The past is split into two forms
according to three paradoxes of time that Deleuze draws from Bergson's *Matter and Memory*. While these paradoxes are beyond the
scope of this essay, they are summarily discussed in Deleuze's description of remembrance. In remembering a present that has already passed, I call it forth into my present moment. In this way,
memory brings together a former present and a present present,
to use Deleuze's vocabulary. However, in this process of remembrance, a former present is extracted from the past in general in
order to be brought to bear on the present present. It is for this
reason that Deleuze will say the past in general (or "pure past") is
trapped between two presents.[27]

In the passive synthesis of memory, one must distinguish between two aspects of the past: particularity and generality. The
past in particular is the former present that is recalled into the

26. Deleuze, *Difference and Repetition*, 77.
27. Ibid., 81.

present present, while the past in general is the accumulation of all former presents that may be possibly recalled in a present present. With each passing contraction, the past in general grows. This growth allows for ever increasing possibilities in regards to what may become the past in particular. This process engenders two kinds of repetition.

The first kind of repetition in the second passive synthesis is the repetition of elements. This repetition denotes the process of contracting a former present with a present present, which occurs in every contraction on a passive level even if this is not made explicit in an active synthesis of memory.[28] The second kind refers to the passage from a present present (which belongs to a contraction with the former present) to the past in general. Above, I described how the two kinds of repetition in the first passive synthesis relied on difference. Deleuze argues that the same repetition of difference occurs in the second passive synthesis. On one hand, the first repetition relies on the difference between the elements of a former present and a present present. On the other hand, the second repetition manifested the difference between the various levels of former presents within the pure past as a whole. It is the differences engendered by these repetitions and the difference between them as such that constitutes the second passive synthesis.

Insofar as the second passive synthesis helped explain a problem with the first (where do contractions go?), it takes ontological priority over the first. The second passive synthesis is required for the first to exist. In other words, there is no habit without memory. Deleuze clarifies the complex relation of the syntheses when

---

28. In fact, Deleuze never tires of reminding us that these descriptions all refer to passive syntheses. ibid., 84.

he says that (the active synthesis of) memory is *founded* on the first passive synthesis, but is *grounded* in the second. Yet on what ground does the second synthesis rest? Moreover, is temporality exhausted with only two dimensions (past and present)? Deleuze answers these questions with his third and final passive synthesis.

The explication of the third passive synthesis begins with a consolidation of the previous two. It is true that the first two passive syntheses constitute time in its preliminary dimensions of past and present, but it is not clear *where* this time is supposed to play out. To answer this question, Deleuze turns to Kant's notion of the transcendental. In many ways, the transcendental is like what has been described as the virtual: it is real, but not actual. For Kant, transcendentals were the "conditions of possibility" that determined what phenomena were open to the experience of a subject. The great contribution of transcendental philosophy, however, was the introduction of time into thought as such.[29] Let's unpack this claim.

Deleuze situates the weight of this contribution around the difference between the Cartesian and Kantian cogito. Descartes had famously argued, "I think, therefore I am" (*cogito ergo sum*). Yet Kant did not believe that a determination ("I think") could imply something that is undetermined ("I am"). In other words, no matter how much I may think, I will never come to a thought or sensation that specifies my existence (as "I am" or selfhood). Kant's solution to this problem is the creation of a third instance between determined and undetermined: the determinable. The self ("I am") neither exists as determined nor undetermined substance. It is rather the "I think" (or empirical self) that is determined from elsewhere. This elsewhere is none other than the transcendental

29. Deleuze, *Difference and Repetition*, 87.

form of subjectivity. The "transcendental I" is outside of thought and is experienced by the passive self as something which acts "upon it but not by it."[30]

Determination necessarily takes place over time. It is in this way that the passive self has time and experiences itself as passing through time. If it is the transcendental I which progressively determines the passive self, then the I must be outside of the time which it determines. It exists in a "pure and empty form of time," but is also able to determine the time of inner sense; for this reason, Deleuze refers to the I as "fractured."[31] Later on, Deleuze explains that the empty form of time is a "form of change [which] does not change."[32] He also describes the transition from the empirical to the transcendental as that from the cardinal to the ordinal.[33] The task of the third passive synthesis is to explain how this works.

The transcendental exists in a time that is out of joint, which means that it is not inscribed in the traditional passage of time. It is what allows time to unfold, rather than letting things unfold in time. This is the sense that the transcendental as ordinal time must be understood: the unfolding of the order of time into past, present, and future. Nevertheless, the past and future that are unfolded from time always exist in inverse proportion to each other. The future is essentially "more" than the past since it presents infinite

30. Ibid., 86.
31. Ibid.
32. Ibid., 89.
33. Ibid., 88.

possibilities.[34]

Despite the inverse proportionality of the dimensions of time, Deleuze suggests some symbols that might stand for it. He is fond of adopting Hamlet's formulation as "time out of joint." Other instructive symbols are the death of God and the explosion of the sun. At issue in each of these symbols is the end of time as we know it. For instance, if the sun were to explode, the very basis by which we keep time would be lost. There would no longer be any discernable past, present, or future, even though the possibility for this unfolded form of time would still be there. This example provides a symbolic representation of what it means for the empirical sense of time to be grounded in a groundless unfolding of time.

The totality of time described just now is the condition of possibility for a temporal series. The series is the *order* of time as an empirical self would experience it. As we have already noted, the future is given in excess. Future actions get imagined as "too big for me," even if they never come to fruition.[35] Deleuze claims that this way of approach future actions is oriented by past possibilities. As long as we cannot anticipate the execution or outcome of a future action based on previous ones, it always seems beyond our reach or "too big." Deleuze refers to Hamlet's indecision to return to Denmark as an example of this orientation in the past. Yet a moment comes when this hesitation dissipates. Hamlet *must* return to Denmark.

---

34. We have already seen how the past is constituted by a finite number of past contractions. Deleuze had argued for the "infinity" of the past insofar as there are an infinite number of former present and present present combinations. Yet this infinity is different in kind than the one discussed in the third passive synthesis. The former regards unlimited substitutions within a closed system, while the latter poses the infinite as such.

35. Deleuze, *Difference and Repetition*, 89.

To carry the example over, Hamlet is transformed in his return to Denmark. He is now on a voyage of revenge. In this way, the act that was seen as "too big for me" is equalized. As soon as the past-oriented self departs toward the future, its present becomes a metamorphosis.[36] This metamorphosis doubles the self, Deleuze says. In becoming-equal to an act that was "too big for me," I am split between my past-oriented self and my present self, between the excessive act and equal act. The act brings about the future, but in doing so it also changes the conditions of excess.

What was formerly "too big for me" is now a part of my past-oriented self. The infinite possibilities of a new future reassert themselves. Every new act seems, again, "too big for me." In this way, Deleuze describes a "secret coherence" between the event and act which necessarily excludes the self; they turn against the self and "smash is to pieces."[37] In its very becoming-equal, the self moves toward the unequal future. Its equality to the act occurs simultaneously with the advent of a new future. But this advent does not allow the self to co-exist with its future. The future escapes into ever new possibilities which remain just as "big" as before: "what the self has become equal to is the unequal in itself."[38]

As may already be apparent, the third passive synthesis brings together the first and second. It constitutes and is constituted by the passage of contractions in the first two passive syntheses. Not only does it complete the account of time with the third dimension of futurity, but it also provides a ground (which is itself groundless) for the production of new contractions. Without new contractions,

36. Ibid.
37. Ibid.
38. Ibid., 90.

habit and memory would not exist. The final synthesis repeats the previous two within a new contraction. Moreover, it repeats the future of infinite possibilities. The double repetition effected by the third passive synthesis reveals the radical implications of Deleuze's reconceptualization of repetition.

The section on the third passive synthesis concludes with the declaration that "all is repetition in the temporal series." The discussion of the first two passive syntheses demonstrated that their genetic principle was repetition. The third synthesis repeats those repetitions while introducing its own unique repetition. Deleuze calls this new repetition the "repetition of the future as eternal return." The future is the promise of the return of the passive syntheses and their redeployment within a new contraction. The future, as the apprehension of novelty which shatters our former self and as the introduction of new possibilities, is essentially the production of "the absolutely new itself."[39] As the repetition of repetition in itself (in the form of the first two syntheses), the eternal return finally reveals repetition *for itself*, which has complete novelty for its essence.

Deleuze stresses that repetition for itself (which is a repetition of differences) takes priority over repetition of the same.[40] Repetition is not a concept for reflection. Deleuze believes that historians treat repetition in this way when searching for historical correspondences between past and present. This view fails to recognize that the past

---

39. Deleuze, *Difference and Repetition*, 90.
40. In fact, it is only with repetition for itself that bare repetition can exist. This part of Deleuze's philosophy, while very interesting, is beyond the scope of this work.

and present are repetitions in themselves.[41] The constitution of the
past and the metamorphosis of the present are repetitions that can
exist only on the condition that we repeat and that what is repeated
is repetition itself (i.e. novelty). The repetition of the excess of the
future disintegrates the default position of the self and its attempt
to become equal to its act. In this way, the autonomous product
of the third passive synthesis returns neither agent nor condition
from the previous contraction. It returns only repetition for itself,
eternally.

The notion of "eternal recurrence" was popularized by Friedrich
Nietzsche, whom Deleuze draws from. Many, however, interpret
Nietzschean repetition in terms of the circularity of time. The
end of time will meet at its beginning and then everything will be
repeated just as it was before. This circle of time repeats only
the same thing over and over. Such would be the case if time
was never out of joint (as the Cartesian cogito asserts). However,
three passive syntheses required the existence of a transcendental
I that is out of joint with time. This transcendental I is fractured
so that time can unfold for the empirical self and be constituted
by passive syntheses. Repetition for itself, which allows time to
continue unfolding, breaks the circle and forces a new circle to

41. Deleuze articulates repetition as action (in contrast to reflection) with the
following example: "It is not the historian's reflection which demonstrates a
resemblance between Luther and Paul, between the Revolution of 1789 and the
Roman Republic, etc. Rather it is in the first place for themselves that the rev-
olutionaries are determined to lead their lives as 'resuscitated Romans', before
becoming capable of the act which they have begun by repeating in the mode
of a proper past, therefore under conditions such that they necessarily identify
with a figure from the historical past." Deleuze, *Difference and Repetition*, 90.

spiral out of itself.[42]

The three passive syntheses have presented a new conception of repetition that does not rely on identity or resemblance and is intimately connected with diachronicity. To each synthesis belongs two kinds of repetition. I introduced these with the terms "bare" and "masked," but Deleuze attributes much more to this distinction: "One is bare, the other clothed; one is repetition of parts, the other of the whole; one involves succession, the other coexistence; one is actual, the other virtual; one is horizontal, the other vertical."[43] Like Barthes, Deleuze acknowledges a dimension of verticality, which is engendered by repetition for itself. Acknowledging this repetition beneath the surface of the actual alters our understanding of the future. Early in the chapter, Deleuze announces that the future is no longer approached in anticipation, but in prediction.[44] As the ground for the active synthesis of experience, the passive syntheses present us with a future that is determinable. This progressive determinability does not lead to a final, necessary determination, but to continuous novelty. Hence, what is predicted is not an entity or event itself, but the set of available possibilities in the actual. More will be said on this below.

What Deleuze has introduced is a kind of transcendental deduction of the passive syntheses based on actual experience. In order to make the predictions of passive syntheses explicit, he requires a methodology that can analyze the virtual from the point of view of the actual. He calls this methodology "transcendental empiricism,"

---

42. "The order of time has broken the circle of the Same and arranged time in a series only in order to re-form a circle of the Other at the end of the series." Deleuze, *Difference and Repetition*, 91.
43. Ibid., 84.
44. Ibid., 71.

which appears as a kind of structuralism of the virtual. In the following section, I will explain how Deleuze maps out the virtual and
its relation to the three passive syntheses. Afterwards, Deleuze will
be compared and contrasted with Barthes and Derrida. From there
it will be possible to extract the notion of repetition for itself and
put it in the service of a new form of structuralism.

Transcendental Empiricism, or The Science of Repetition

Everything depends on intensities. It is at the end of the third
passive synthesis, when thought and prediction emerge, that a concept of intensity can be formulated. Although it would be wrong
to understand this concept in an entirely Simondonian sense, this
articulation is certainly operating in the background of Deleuze's
discussion. Yet for Deleuze, the intensity is not some kind of metaphysical power or physical force: it is difference itself. He explains:
"The expression 'difference of intensity' is a tautology. Intensity is
the form of difference in so far as this is the reason of the sensible.
Every intensity is differential, by itself a difference. Every inten-
sity is $E - E'$, where E itself refers to an $e - e'$, and e to $\varepsilon - \varepsilon'$
etc."[45] Hence, it is the differences discovered by the three passive
syntheses that become expressed as intensities.

Intensities, however, as bundles of differences, are also responsible for what appears. Deleuze will also call intensity the "sufficient
reason of all phenomena."[46] To unpack this a bit, we must return to
the situation prior to the syntheses of time. At first, there was only
a passive, empirical subject that was able to experience immediate
sensations. These phenomena first appeared as crude impressions.

45. Ibid., 222.
46. Ibid.

Once the three passive syntheses have worked over a multiplicity of impressions, their context comes to the fore. This context is none other than a system of repeatable differences or, in other words, the virtual. In this way, the syntheses do not create differences between immediate sensations, but draw off the differences that are already latent in such sensations. Each sensation is always already an intensity.

Perhaps now it will become clear why Deleuze had referred to intensity as the "reason of the sensible." The differences brought together in the passive syntheses are the condition for the active syntheses of imagination, memory, and thought. The passive syntheses generate, via intensity, sensible experience. Yet the sensible individual is not what is primary in generation. The individual only exists insofar as it stands out of the matrix of differences from which it emerges. Such standing out is what Deleuze calls "actualization." Understanding actualization as the passage from insensible virtuality to sensible actuality (i.e. the generation of experience) is the first step of transcendental empiricism.

When he introduces the idea, Deleuze calls transcendental empiricism the "science of the sensible."[47] Its goal is to work backwards from the sensible in order to understand how it is generated. One should notice that the question concerns the "how," not the "what." For Deleuze, it is senseless to interrogate the quiddity or what-ness of the virtual. The virtual is composed of intensities (i.e. differences). If one were to ask what the difference is of, the answer would inevitably be: difference of differences. Thus, transcendental empiricism does not investigate entities, but the conditions of real experience. In doing so, it takes the sensible as its primary

47. Deleuze, *Difference and Repetition*, 56.

object. Yet, this is not the "sensible" that has been the object of so many philosophers in the past; it is the sensible insofar as its identity is "swallowed up in difference."[48] The "sufficient reason" that transcendental empiricism searches for behind appearances is a "'strange' reason, that of the multiple, chaos and difference."[49]

The account of the three passive syntheses above helps us understand how real experience is taken up as an object of analysis in transcendental empiricism. This object, unlike many other scientific objects, is a four-fold moment of difference composed of selection, repetition, ungrounding, and the question—problem complex.[50] It is clear that selection, repetition, and ungrounding correspond to the three passive syntheses, respectively. The last aspect that is constitutive of actual experience is the verso of the actual, i.e., the virtual. I have already said that intensity is recognized for what it is after the third passive synthesis. Along with it, the virtual is produced for the first time. In many ways, the virtual is implied in intensity. If intensity is the expression or individuation of a difference, it establishes a fundamental difference between the realm of individuation and that of difference in itself. The virtual is then understood as what directs and determines intensities that take up specific forms through the process of individuation. The virtual, like the *langue* of traditional structuralism, is thus the focus of transcendental empiricism. I will briefly explain the structure of the virtual before concluding what we have gained from Deleuze.

The virtual is structurally organized on two levels. First, it is broken up into a multiplicity of structures called Ideas. Second, an

48. Ibid.
49. Ibid., 57.
50. Ibid., 68.

Idea itself is a structure which contains various elements: differential field, differential relations, and singularities. The differential field is the "space" of the Idea.[51] It is essentially the limits of the intensities that belong to that Idea. The differential relations are what exist between singularities. Each singularity is a point within the differential field of the Idea and the differential relations are the ways these singularities interact with each other. Individuation occurs when differences are gathered together in a synthesis. The Idea provides structure for intensities, which determine what is individuated. Hughes rightly argues that the relationship between Ideas and intensity can be understood as that between DNA and cytoplasm.[52]

To get a clearer picture of just how individuation works, we can turn to John Protevi's example of American football.[53] In American football, each player would represent a singularity that has a differential relation to each other player. So, the center has a certain relationship with the quarterback that the wide-receiver does not. The differential field mirrors the actual field in certain repsects. In this example, the elements of the Idea are clarified, but what is interesting about it is how the pieces interact. Intensities propel the game in a certain direction and various thresholds exist that produce changes in intensities (in this case: points). So, the obvious thresholds are the first down line, the goal line, and the uprights. Crossing the goal line produces six points and resets the relation of the singularities (e.g. the other team gets the ball after the extra

51. Space is a quality of the actual. So, when discussing the topology of the virtual, I am using "space" as a metaphor.
52. Joe Hughes, *Deleuze's Difference and Repetition: A Reader's Guide* (New York, NY: Continuum, 2009), 170.
53. John Protevi, *Political Affect: Connecting the Social and the Somatic* (Minneapolis: University of Minnesota Press, 2009), 13–4.

point). Protevi also asks us to imagine tweaking the Idea slightly by changing the differential relations or field. In this example, if you take away some differential relations (the rules of American football) and alter some singularities (make the ball oblong), you would have the Idea of rugby. It should be remembered that each Idea is a structure of the virtual. Each American football game that is played represents an actualization of that Idea. So, if we want to understand this in orthodox structuralist terms: the Idea is the *langue* and each individual game is *parole*.

Indeed, it is becoming exceedingly helpful to articulate Deleuze in structuralist terms. Does this mean that transcendental empiricism is just a variety of structuralism? No, because there is a crucial difference in methodology. The whole issue revolves around traditional structuralism's desire for synchronicity. Saussure's *langue* was stable and timeless; Deleuze's virtual is meta-stable and dependent upon time. Given that these different methodologies have defined structure in incompatible ways, their approaches have likewise been at odds. Saussure had to fix every sign within a certain temporally restricted syntagm and paradigm; Deleuze claims that the "principle of transcendental empiricism" is genitality.[54] Contrary to traditional structuralism, transcendental empiricism treats its *langue* like an open question that is constantly unfolding and will never be answered with any finality.[55]

Undoubtedly, transcendental empiricism begins from a fundamentally different premise than traditional structuralism. The great

---

54. Deleuze, *Difference and Repetition*, 147-8 and 176.
55. Deleuze, *Difference and Repetition*, 192; Deleuze will make a very similar argument against structuralism in his work with Guattari. Gilles Deleuze and Félix Guattari, *A Thousand Plateaus: Capitalism and Schizophrenia*, trans. Brian Massumi (Minneapolis: University of Minnesota Press, 2009), 237-9.

contribution of Deleuze's perversion of the structuralist method is
that he makes diachronicity essential to structure and structuralism. We can return to Protevi's football example to see why this
would be useful. Football's structure/Idea (its rules and modes of
articulation) is continually changing. For instance, new strategies
for scoring alter how the game is played. For a while, quarter backs
did not have the well-developed arm strength required for throw-
ing the ball accurately at over forty yards. With this development,
defensive formations needed to adjust to a more invasive passing
strategy. New possibilities and potential outcomes were created.
These virtual possibilities, as they played out, led to new demands
in the actual. In the case of the evolution of passing, rules needed
to be put in place to protect the integrity of the game and its play-
ers (e.g. roughing the passer, passing interference, etc.). One can
also consider the recent addition of the trapezoid restriction area
behind the net in National Hockey League games. Technological
advancement, which led to the increased mobility of goaltenders,
spurred mutations within the differential field. (More will be said
on how this is possible below.) This process reveals the infinite development of the virtual, whose transformations can be predicted
with transcendental empiricism.

Deleuze's overturning of structuralism is not merely a preference for the diachronic over the synchronic. All of this is possible
only through his concept of repetition for itself. Of course, pointing
out the diachronicity of structures is not a novel claim. As we have
seen, Barthes had already done that. In his case, however, repetition became the enemy of meaning. It was supposedly responsible
for coagulating structures and solidifying meanings. Deleuze lib-
erates repetition and honors it as the fundamental movement of

all diachronicity. Yet, it is only through his reconceptualization of repetition that he is able to link it to meaning-production. Repetition, instead of being the morgue of signifieds, is the factory of signification.

To understand how meaning is produced and transformed by repetition, let us consider two examples. First, a door-to-door sales-man returns to a certain house twice.[56] He first goes to the house with the intent to sell some product. By the end of the day, he is exhausted and does not remember going to that house, so he re-turns. The bare repetition is being at the house again; the masked repetition is forgetful and returns to the house in a different way and with a different meaning. Second, repetition for itself has been illustrated in a passage from Virginia Woolf's *To the Lighthouse*:

> What was the good of doing it then, and she heard some voice saying she couldn't paint, say-ing she couldn't create, as if she were caught up in one of those habitual currents which after a certain time forms experience in the mind, so that one repeats words without being aware any longer who originally spoke them.[57]

In this case, it is not the origin that is eclipsed through repeti-tion (as Derrida would suggest), but the origin that is contingently established through various repetitions. The character from *To the Lighthouse* can only hear the repetition of the words, which ef-face their origin. Repetition itself becomes an origin that infinitely reproduces itself. Repetition, then, opens the very possibility of repression and memory.

56. I am indebted to Jake Hamilton for this example.
57. Virginia Woolf, *To the Lighthouse* (New York, NY: Oxford University Press, 2008), 131.

In one of the densest passage of *Difference and Repetition*, Deleuze reverses Freud's famous repression hypothesis: "We do not repeat because we repress, we repress because we repeat."[58] In other words, forgetting (or repression) is intimately related to the process of repetition. One must forget in order to remember or repeat. The salesman forgets that he had already visited a certain house, opening the possibility for repetition. Repetition recalls a former singularity, but brings it into a new milieu. Repetition is remembrance, albeit a special (Proustian) kind.

As a process of forgetting-remembering, repetition also becomes a movement of question-response between the virtual and actual. Deleuze describes the Idea as a problematic structure which consists of chaotic, heterogeneous series of intensities. Think, for example, of the various, implicit possibilities of an embryo. When these series connect and resonate, a "solution" to the Idea is individuated. So, when the intensities in the reptile embryo resonate at a certain frequency because of the temperature of the environment, its gender is selected. Each individuation of the Idea is a repetition of its structure, but is different each time. In the same way, each instance of a football game follows the same rules, but no two instances are ever exactly alike. The play between the problematic Idea and the situated response reveals an implicit meaningfulness. Significance is literally *produced* by these structures in how they affect (and are affected by) different situations. As the Idea becomes actualized, it changes the conditions of the actual, which in turn alter the intensive potentials hidden in the virtual. Deleuze calls this back-and-forth motion *vice-diction*.

Vice-diction is opposed to Hegelian contra-diction. On one

58. Deleuze, *Difference and Repetition*, 105.

hand, Deleuze argues that Hegel's dialectic subordinates difference to the law of identity. Dialectical contradiction can only imagine the cancellation of polar opposites. Actualization, though, is rarely the sublimation of an ideal binary. The relation between the actual and virtual is multifaceted and embodied by a multiplicity of intensities. Vice-diction, on the other hand, is the quintessential problem-solving process for this dynamic. Each response in the actual creates a new problematic in the Idea. To explore this operation, Deleuze uses the example of a swimmer.

In this complex example, Deleuze links repetition, difference, and learning within the context of a body swimming through waves:

> When a body combines some of its own distinctive points with those of a wave, it espouses the principle of a repetition which is no longer that of the Same, but involves the Other – involves difference, from one wave and one gesture to another, and carries that difference through the repetitive space thereby constituted. To learn is indeed to constitute this space of encounter with signs, in which the distinctive points renew themselves in each other, and repetition take shape while disguising itself.[59]

Each wave presents a new challenge to the swimming, but her ability to interpret and respond to the challenge is what keeps her afloat. A stroke is only individuated if it keeps the swimmer afloat by responding to the distinctiveness of each wave. In this way, the stroke is repeated, but differently than the stroke before it. The stroke must respond to the shifting requirements for buoyancy. The swimmer is a symbol of vice-diction because she is able to tarry

59. Ibid., 23.

with the virtual in a way that is immanently meaningful. Here, the delicate nature of meaning, which Derrida attempted to exploit, is fully appreciated by Deleuze. That is, it only takes a moment for a swimmer to become a corpse and meaning to fall into oblivion.

The definition of learning given above is crucial to understanding the practice of transcendental empiricism. All learning involves an "encounter with signs." Deleuze argues that this means doing is an essential part of learning. Instead of the classical teacher representing a maneuver to a student and the student reproducing the action, learning should take the form of a sensitivity to signs and an engaged response to them. Hence, Deleuze says the ideal teacher is one who says, "do with me," instead of, "do as I do."[60] In this sense, the practice of structuralism is democratized. It is no longer solely the linguistics professor that practices the methodology, but also the doctor, mechanic, lover, etc.[61] The farmer is a transcendental empiricist when he finds red dock (*Rumex bucephalophorus*) and realizes that his field is calcium deficient. Structuralism is no longer just a way of recording dead structures, but of reading virtual transformations in a way that allows one to have a meaningful response to them.

Deleuze has presented a powerful response to Derrida's claim that meaning is always differed. Yet the two theorists also agree on a number of points. Deleuze's repetition for itself ends up looking a lot like Derrida's eternal play of signifiers. The key difference between the two is that Deleuze fractures reality into virtuality and

---

60. Deleuze, *Difference and Repetition*, 22–3.
61. Deleuze refers to this kind of encounter with signs as an "apprenticeship." The apprenticeship to signs plays a significant role in his work on Proust, Gilles Deleuze, *Proust and Signs: The Complete Text*, trans. Richard Howard (Minneapolis: University of Minnesota Press, 2000), 26–38.

actuality. Derrida's ruminations are stuck in a pseudo-virtual zone of differentiation. Signifiers can be substituted eternally because they never take up a signified within the actual, whether necessarily or contingently. The type of structuralism that Derrida is attacking wants to claim that signifiers are necessarily tied to a signified at a given moment, although which signified is selected can change over time. Deleuze provides a middle path between the two extremes by suggesting that signs are contingent manifestations based on the possibilities determined by virtual problematics. The actual-virtual distinction allows Deleuze to formulate an exchange that produces local, contingent meanings, while admitting the existence of irreducible difference. Derrida remains trapped in the virtual and does not realize that one does not need an absolute origin to act as if an origin had existed. Deleuze and Derrida would agree that there are no absolute meanings (in structures or otherwise), but Deleuze's vice-diction allows for the local production of meaning within structures. The play of meanings, rather than the disavowal of meaning as such, is what makes Deleuze's thought more akin to Barthes' than Derrida's.

Deleuze and Barthes are both concerned with eggs. Deleuze says the world is an egg; Barthes says "Sarrasine" is an egg-text.[62] What this means is that out of these small differential fields, limitless possibilities abound. An embryo or text can produce infinite phenotypic interpretations. Thus, rather than waiting for the eggs to hatch, Deleuze and Barthes each set out to construct a general embryology of structures. In fact, many of Barthes' concepts have direct correlates to Deleuze's thought: secret and repetition, history and second passive synthesis, field and Idea. Deleuze, however,

---

62. Deleuze, *Difference and Repetition*, 216; Barthes, *S/Z*, 200.

expands on Barthes' work. Barthes' radical insight into the verticality and secrecy of texts is rehabilitated in a kind of repetition
that is both "vertical" and "secret."[63] Moreover, the forgetfulness
at the heart of Barthesian meaning-production becomes the key to
comprehending the movement of repetition for itself.[64] Deleuze and
Barthes are the forefathers to a new kind of structuralism: hyphology. Yet one must take careful steps away from both philosophers
to grasp this project and adequately respond to Derrida.

In *Difference and Repetition*, Deleuze replaces the concept of
the subject with that of the "larval subject." The difference is that a
larval subject is a meta-stable organism that is constantly changing
and individuating in each new moment; it does not have an identity
or essence. The larval subject is a fractured-I, lacking the coherence
of the faculties, and allowing Ideas to swarm through it like tiny
ants.[65] It is important to realize that structures have the same
nature. At this point it is possible to move from *larval subjects* to
larval structures. Larval structures are the object of hyphology.

Before proceeding it would be wise to summarize the progress
of this essay so far. We began with reviewing Derrida's critiques
of structuralism and asserting that some were more invidious than
others. To assess the strength of Derrida's arguments, we turned to
Barthes' practical engagement with structuralism. Barthes showed
us that structuralism is more elastic than Derrida had conceived,
but there was still an issue of how meaning could be produced. It
was here that the concern over repetition was raised. Repetition
was responsible for closing structures in Saussure and Barthes, so

63. Deleuze, *Difference and Repetition*, 18 and 289.
64. Ibid., 7–8, 16, and 109.
65. Ibid., 277.

we turned toward phenomenology and Deleuze to assess the role of repetition. Deleuze has equipped us with a new understanding of repetition and the nature of meaning. This was where we ended. The goal remains to formulate an adequate response to Derrida's critiques. So far, we have seen many useful tools come from Barthes and Deleuze, but there is still not a method. The rest of this essay will be dedicated to developing a program for the hyphologist and putting it into practice. The few deconstructive critiques that have not been responded to will be accommodated by hyphology.

5

Hyphology

The first concern of this section is to define the topic of hyphology. It must, at all costs, be differentiated from structuralism and post-structuralism. Hyphology grows out of the tension between these two positions. From structuralism, it inherits the concern for meaning-production and the emphasis on the structure; from post-structuralism, it adopts the incompleteness of the structure and the concern for the diachronic. Everything else within these traditions (strategies, dogmas, etc.) is left to the abyss.

Superficially, hyphology may appear the same as transcendental empiricism. While these two methods have much in common, they differ on some critical issues. One of the most pressing is that Deleuze does not offer a robust theory of signs. His distinction between "natural" and "artificial" signs fails to address the matrix of significant differences that give rise to the sign.[1] In place of this fundamental distinction, the hyphologist privileges the Barthesian notion of *weight*. The concept of the weight of a sign not only evades the oversimplification made by the transcendental empiricist, but also eludes the Derridian accusation of logocentrism insofar as the "natural" order of signs refers to that which is present.

By all intents, hyphology is a scientific method. Yet it must be distinguished from the scientific method proper. The goal of the scientific method is to narrow down an environment to one variable so that it can determine the elasticity of that variable. However, as soon as one removes that variable from its context, it no longer operates in the same way. *A thing is only as meaningful as its repetitions.* That is to say, a thing is only as meaningful as its differenc/tiations (how it is distinguished from other things and from itself).[2] Hence, what hyphology proposes is that a variable be analyzed within its context, despite the contingency our analysis must accept. This move allows us to escape the invidious closure of structure that Derrida warned about while also maintaining, in part, the goals of science.

In terms of its positive project, hyphology has rebuilt itself from the most basic concept that attends structure-formation: repetition. The goal of hyphology, as a middle path, is to produce mean-

1. Deleuze, *Difference and Repetition*, 77.
2. Of course, removing something from its context in the hopes of repeating it leads us back to all the trappings of Derrida's deconstruction of communication. Cf. "Signature Event Context," in *Margins of Philosophy*.

ingful statements about structures while recognizing their contingency. Structuralism was only concerned with necessary conditions derived from synchronic analysis; post-structuralism negated
the possibility of any meaning whatsoever. The detour through
Deleuze revealed the point that *meaning is real.* In life, faced with
constant problematics, one must decide whether one wants to swim
or drown. How meaning can actually be produced will be of the
utmost concern.

The trace engenders a process of contingent signification similar to the question-response format.[3] Spacing allows for this to
occur: the signifier asks and the "signified" responds. Derrida argued that this arrangement made it so that meaning was never
present. However, meaning continues to be produced everywhere
and at all times, whether the signified is reached or not. In this way,
Derrida confuses the lack of the signified with a lack of meaning.
The crisis of meaning is amplified given that signification requires
the sign, and also every-thing is a sign.[4]

Yet the closure of signification and its metonymic skid (within
the absolute reading, the perfect stroke, or other final assessment)
effaces not only the *contingency* of the meaning, but also its *meaningfulness.* Due to this myopia, hyphology is concerned with illuminating the secret (i.e. the very locus of meaning), which is a black
body radiator.[5] The impossibility of such an illumination is approached repeatedly; only by re-reading can one uncover the play of
codes underneath the structure. Thus, like a game of twenty questions, the hyphologist unfolds the structure by attempting (though

3. Derrida, *Of Grammatology*, 65.
4. Ibid., 50.
5. This means that the secret attracts all light and hides it within itself. All
the light in the universe would not illuminate the secret.

never succeeding) to exhaust its possibilities through responding to its local problematics. The reason this process can never be completed is that the limits, if one could even apprehend them, of the structure are beyond its present horizontality and horizon.

Hyphology is concerned with verticality. The vertical only appears in the diachronic or secretive. In this way, one must understand repetition to be more than the basis of structure-formation: it is also that movement which the hyphologist investigates. Difference and repetition are the foundations of verticality insofar as the sign differs from itself through difference-in-itself and repetition-for-itself. This fact is why Deleuze says that repetition is a "secret verticality."[6] Signs must be tracked through elaborate echographies in order to determine their immanent value within a structure. Every encounter between structure and structuralist, between object and subject, implies two vertical spires, like Hollywood premier lights forming immaterial intersections.

Barthes presents a profound example of verticality in his confrontation with the Latin grammar puzzle. The meaning of the Latin grammar example is only clear within a certain pedagogical context where some Latin has been learned in the past. At this juncture, there are numerous potential configurations of Latin projected into the future. The Latin grammar example from Barthes is a sign that interfaces with the student: both of which contain infinitudes of diachrony within them. Hence, hyphology requires a new conception of the text that facilitates a method of analysis that is no longer merely subjective or objective.

It is clear that structuralism and post-structuralism begin with pre-determined conceptions of the text and its limitations. Struc-

6. Deleuze, *Difference and Repetition*, 18.

turalism explores the orange-text while it peels away the skin to trace the veins; post-structuralism pursues the onion-text through various transparent layers that eventually lead to an absence of origin. Hyphology, in contrast, treats the text as a plastic loofah. The loofah is made up of layers and nothing more. The layers are mostly transparent, but rough when rubbed against the skin. As one peels back each layer, one eventually finds nothing at the center. The great comedy of the text is revealed in the reader's feeling of loss at the non-existence of a center that holds the loofah-text together. Yet the reader is unable to deny the reality of each layer obscuring the non-center. The loofah-myth is a nebula of significations (rubbing each other and carving out) without a central signified. Although it lacks a center, the "center" is felt with each lathering. As soon as the loofah-text is dismantled, it becomes a horizontal chain of insubstantial fabric. This mutilated text provides the perfect object of analysis for classical structuralism, but it is no longer able *to do* anything. Even the most routine activity (e.g. showering) is implicated in hyphology.

It would be impossible to conclude at this point, like so many dense treatises that leave the reader confused and agitated. Hyphology is not a theory, but a practice. To demonstrate this fact, I will now take up two mythologies, which are explored through the lens of hyphology.

6

Two Hyphologies

The following two hyphologies are not meant to be a complete analysis of the given structure, but only an attempt to provide some illuminating insights. The previous section hints that this is all hyphology can hope for, but it would be false to assume that this entails the inability of saying anything meaningful about a structure. The hyphologist studies repetitions just like the mythologist, which implies that the hyphologist never searches for the origin. Although it is popular to speak of origin (*arche*) in philosophy, it must be remembered that the origin is always contextualized as that thing which produced the current situation. Hence, it is the local significations that are paramount in these hyphologies.

Automobile Advertising

There is a unique phenomenon that is peculiar to the advertising of automobiles. Commercials portray their product in motion. Cars drive through fields, deserts, mountains, and cities, which are all completely empty. Even in the cities, there is not a single car or person around. Trucks are magically loaded with heavy objects to show off their suspension, but there is no one doing the loading. Even the driver is often totally occulted. When the driver is revealed, it is usually a masked professional or only the hands. The question arises: why is the automobile always presented in isolated environments? What can a synchronic analysis tell us?

To really understand the commercial-structure, it must be isolated in time from any transformations that may have existed before it or will exist after it. As Barthes instructs the reader on the last page of *Elements of Semiology*, this method is infinitely favorable to one that accounts for the evolutionary development of the commercials. (This is also a condemnation of the trace.) So, one is left with a bundle of commercials that share the characteristics described above. A contrast exists between the product and its environment that resembles the figure-ground relation. The car is emphasized by being in an empty field; it draws the viewer's attention like the eyes of the Mona Lisa. The absence of the driver is an *invitation* to imagination. (To *invite* the viewer to the dealership has become common practice. Buying a car is treated like a festival, which should remind the reader of the lyric from the Claude Channes song in *La Chinoise* [1968]: "revolution is not a

banquet.") One can picture oneself steering the car through the
vacant city streets. The commercial allows the viewer to test drive
the car in the very best environments. Thus, a synchronic analysis
can provide an answer to our question. It has revealed the viewer's
relation to the commercial to be part of the very sign-system established by that commercial, and it has also explained the meaning of
the isolated environment to be a tactic of emphasizing the product.
The quality of this answer will now be weighed against that of a
hyphological analysis.

One cannot underestimate the critical role played by diachronic
elements. Even in the last page of *Elements of Semiology* there
is a tension between completely negating these elements and for-
mulating a method that gravitates around them. To interpret the
commercial, the past of the automobile and its marketing must be
summoned. This means that an exploration of the verticality of
the signs in the commercial must be conducted. When watching or
using a car, we are participating in a mnemotechnology that carries
its history within it. To understand the commercial today, we will
explore the commercials of the past.

As soon as a market for the car developed, it became a recre-
ational object. The car was designed for transportation, but adver-
tised for recreation. This can be seen in the advertisements for the
first affordable car: the Ford Model T.[1] The Model T represented
an escape from the city that was not time-consuming or expen-
sive. Car owners could run away from city life for an afternoon and
have a picnic in the country. The car owners were predominantly

---

1. Ford Motor Company, "Freedom for the Woman Who Owns a Ford,"
The Henry Ford, accessed January 14, 2020, http://ophelia.sdsu.edu:
8080/henryford_org/03-23-2014/exhibits/showroom/1908/ads.html.

families. Thus, the Model T extends the private sphere, but it also continues to maintain it at the same time. The car protects the privacy that the family shares in its home and extends that privacy to places even more secluded. By the 1950s, this foundational aspect of the automobile will be proliferated in the forms of drive-in restaurants and movie theaters. Hence, the car represents the border between the private and the public. This realization clarifies the significance of contemporary car commercials in a way that the synchronic analysis could not. It is not the emphasis of the car that the commercial-structure draws attention to, but the car-as-private-space and the distancing of the public sphere. Yet this analysis can be carried even further.

The orthodox structuralist would retort that the history has nothing to do with how commercials are created today. As we have seen, this history helps uncover the peripheral effects on the structure, which should also be considered to be part of it. So, the hyphologist accepts that *there is no outside the text* insofar as this principle takes on its significance found in both Derrida and Barthes. One peripheral effect of the culture of the car commercial is the public phenomenon of a person singing along to music in the car.

Witnessing a person sing along to music in their car is an absurd event. The more invested the person is into the music, the more absurd it becomes. Albert Camus claims that the absurd cannot be explained, it must be encountered. The example he gives in the *Myth of Sisyphus* is not that far from our own: a man on the phone behind a glass partition. However, the goal of hyphology is precisely to explain this phenomenon. Indeed, one must develop an analytic of the absurd. The synchronic analysis cannot speak to this event at all, but the diachronic assessment reveals the origin of the absurdity through the very tension found in the history of the automobile. The absurd, in this case, arises through the paradoxical revealing of the private within the public. The automobile, more than ever before, is a personal space that is isolated from the public, yet it is also revealed to the public in its very formation. Even if one were to sympathize or act similarly, the phenomenon persists in its absurdity because it consists of displaying the private act to the public. This absurdity could not be understood without following the traces of how the automobile has been culturally represented in the past.

The hyphological analysis, which includes diachronic elements, of automobile advertising has provided an adequate response to the question first raised over the commercials. The synchronic analysis gave a consistent answer, but was an ultimately reductive answer that could not explain related phenomena. The mythology of automobile advertising has presented two important points: the necessity of including diachronic elements in structural analyses, and the impossibility of closing a structure due to peripheral effects that create significant alterations within the structure.

Frasier

The television sitcom *Frasier* begins with an invitation. Within the first few episodes, there is a sign that signifies the existence of signs in general: a book on Frasier's shelf by the structuralist Umberto Eco. The show introduces itself like those famous lines from Walt Whitman: perhaps it appears to contradict itself, but it contains multitudes. It soon becomes clear that everything must be interpreted to be a sign or a signifier for something else.

Frasier embodies exactly what Barthes had called *differential geology*. It contains, within itself, its own dynamic of sign relations and meaning-production. This is made clear by Martin's chair, the center of so many episodes. At one point, Frasier, against his will, sits in the chair and finds it fantastic: "When you sit in it you don't have to look at it!"[2] This paradoxical position of fully inhabiting a space yet not perceiving it is where Barthes situates the beginning of structuralism.[3] Martin's chair is the Eiffel Tower. The space is both spectacle and the position of spectatorship, which is where the task of defining the structure begins. It is at this juncture where we are completely immersed in *Frasier* but also trying to stand outside of it, that we will begin our descent into its differential geology.

Like the veins in a block of marble, *Frasier* has rigid lines of demarcation between characters. Each character can be fitted into one of two categories, with one special exception that will be discussed below. The asymptotic thresholds that divide the various intensities emanating from each proper name run parallel with economic boundaries. It is clear that there is a leisure class and a working class of characters. Each sign can be defined to be either *bourgeois* or *proletarian*.[4] This strict division of signs is seen in each episode. For example, Frasier calls attention to it when he says, "For a layperson she has a way of cutting right through the

2. *Frasier* (National Broadcasting Company (NBC), 1993–2004), Season 7, Episode 10.
3. Roland Barthes, *The Eiffel Tower and Other Mythologies*, trans. Richard Howard (Berkeley: University of California Press, 1979), 3–26.
4. I am adopting these names mainly because of the alienation implied between these two economic classes. This is not a Marxist analysis of *Frasier*! Hyphology seeks to uncover the internal dynamics inherit in each structure, which sometimes includes class conflict. The Marxist demands of each structure: "Say *dialectical materialism* when I question you!" The hyphologist scrutinizes each sign like a child at the playground.

crap."[5] The ultimate goal of the show is to explore how people from different generations and cultures could get along. Despite this simple infrastructure, one other category exists, which highlights the division of the first two: *ambiguity*. Very few signs fit in this category, but it is vital to the internal mechanisms of the show. It would now be useful to turn to some examples of these signs.

Each and every object or character in *Frasier* can fit into one of the three categories: bourgeois, proletarian, ambiguous. Frasier's apartment is dense with bourgeois signs. Even its structure consists in signs. The balcony signifies privacy and hierarchy. When the door is closed, nothing can be heard of the outside world and only the politics of the interior become meaningful. The balcony overlooks all of Seattle. (The picture was taken from a mountain where no actual apartment building exists. In this way, it becomes a phantasmatic Panopticon.) Frasier's apartment is a veritable ziggurat, much like the one from *Metropolis*. Even his front door becomes a complex, bourgeois sign. The doors of his apartment are larger than any others in the *Frasier*-world. His front door is the Napoleon monument, through which he allows his esteemed guests to pass when his servant opens it under his watchful eye. The door number reads *1901*, the first year of the 20th century. (Which is also significant for the show as a whole since the 20th century brought several challenges to an ever expanding capitalism, i.e. 1917, 1936, 1968, 1999, etc.). So, Frasier's apartment coincides with the beginning of a new era, not only in his personal life (*Frasier* is a spin-off of *Cheers*), but also in the creation of a new space and time populated by a plentitude of bourgeois signs.

5. *Frasier*, Season 2, Episode 5.

The most interesting proletarian sign would have to be Martin's chair, which is the center of several controversies. It, too, is part of Frasier's apartment, but subsists in its own space. Frasier's apartment is designed so that one must take one of two routes to get from the kitchen to the front door. Unsurprisingly, one is bourgeois and the other proletarian. Martin's chair is situated in a constellation with the television, TV dinner tray, and Eddie, the dog. The chair, as Martin passionately reveals in one episode, carries a history with it. To Frasier it is an old chair that an old man cannot let go of, but to Martin it is the place where he first held his son and where he was when Kennedy was shot. The antique pattern on the chair even matches Martin's daily uniform of plaid. However, the chair is eventually destroyed.[6]

The death of the chair is symbolic and revealing. It is murdered by a tripartite of bourgeois signs: Frasier, the balcony, and the telescope. Frasier pushes the chair, by accident, over the balcony. It falls several stories and is destroyed on impact. Sun shines through the telescope onto the chair and lights it on fire; the corpse is desecrated. Frasier is devastated that he would ruin something so precious to his father. He puts in a lot of time trying to find a replacement, but eventually needs to have one custom-made at a high price. This event uncovers the implicit violence that exists beneath a surface interpretation of *Frasier*. It is not the destruction of the chair, the homicide of proletarian signs, that is violent, but the impossibility of reconciliation. Frasier still does not understand why the chair was significant in a proletarian context and merely attempts a bourgeois solution via spending. Like a couple in an argument, the characters talk *at* each other, rather than *with* each

6. *Frasier*, Season 9, Episode 7.

other. The category that solidifies this alienation is the ambiguous. Every *Frasier* episode begins with the ambiguous. The jazz song that opens each episode is the first ambiguous sign the viewer discovers. Jazz was historically developed by an oppressed class, but has been used since to signify high culture and refined tastes. It straddles the first two categories. One of the only times it appears in the actual show is when Daphne, the other ambiguous sign, is listening to it. Daphne is an ambiguity machine. As a character, she produces uncertainty and fuzziness with regards to the categorization of signs. Immediately before the scene where she listens to jazz, she attempts to enter an argument between two characters, one bourgeois and the other proletarian. She passionate argues for a position, but the other two characters cannot decide who she is agreeing with. This is a major trope of the show and happens several times.

At other times, Daphne sparks confusion with a certain linguistic ambiguity, "Dr. Crane," which could refer to Frasier or Niles, his brother. Despite this common occurrence, there is no attempt made to clarify which Dr. Crane is being named. Daphne has an English accent, which is often taken in an American context to be a sign of intelligence, but she works as the house maid and Martin's physical trainer. She is important as a third category that displays the tension between the first two. There are many signs in the show; some are inert and permanent while others are active and temporary. The way that signs are performed plays a significant role in interpreting how meaning is produced within *Frasier*.

Signs can be performed in three ways: exhibition, imitation, and monumentalization. Exhibition is the most basic way. A character will perform their sign-category in certain ways. For example,

Frasier (bourgeois) obsesses over the placement and care of his
African sculptures (bourgeois). Imitation is the reverse of exhibition: a character will perform a sign from the opposite sign-category
in a sarcastic or ironic tone. For example, Niles (bourgeois) will
sometimes adopt an unusual vernacular (proletarian) to signify he
is doing something unnatural. When Niles has to pump his own
gas, he says he is "learning to be *handy*."[7] The goal of this performance, of course, is to alienate the opposing sign-category by
devaluing it. The final mode of performance is monumentalization, whereby an object takes on a meaning and persists in the
background for the entirety of the show. The key example here is
the telescope, around which a conflict over human rights develops.
The bourgeois characters find it fun and normal to spy on others
from their Panopticon, while Martin (proletarian) believes that it is
morally reprehensible. The signs and their performance sometimes
are guided by a handful of tropes, which continue to concretize the
violence between the sign-categories.

The tropes of *Frasier* often center on ambiguity and alienation.
Miscommunication is present in many of the episodes. Characters will find themselves agreeing to a certain plan, but pursuing
something entirely different than the others because of a misunder-
standing (which is often brilliantly cloaked). Numerous episodes
revolve around the "sharing" of stories, where each character takes
a turn. The issue here is that the others do not appear to listen, but
only ruminate within themselves. This activity is the polar opposite
of another trope where everyone talks *en masse* at each other, but
no one listens. Nothing is shared, but everything is expressed. Re-

---

7. The emphasis on *handy* is key. Manual labor is another proletarian sign
Frasier, Season 2, Episode 14.

occurring jokes and insults round out the continuity of the series. Taking account of the signs, practices, and tropes of *Frasier*, one can draw structural conclusions about it.

A surface reading of *Frasier* would be deceiving. The show progresses to a point where each character resolves their conflicts with the others (who are often of a different sign-category). The message is that, despite their vast differences, the bourgeoisie and proletariat can get along peacefully. The difficulty with this interpretation is that an analysis of the signs, practices, and tropes suggests the very opposite. Characters selfishly talk at and against those of the opposite sign-category without ever coming to reconciliation. Each performance (i.e. signifier) is a hollow echo pointing to a transcendental signified: class struggle. The trace, rather than becoming the impossibility of meaning, represents the production of meaning. Signs and performances become congealed under proper names, which disclose concrete relations between them. In this way, the structure (i.e. *Frasier*) has a past and a future. The diachronic necessity within this reading, insofar as the text is never closed, is clear. Hyphology is a search through the traces for the contingent laws that govern the structure and direct meaning-production.

A question can be raised: What would a text look like in which the signs are reconciled? If signs can never be reconciled, then every text would be based on violence and the implicit meaning of *Frasier* would not be unique. In a sitcom like *Frasier*, the goal is often to form concrete characters that are at odds with other characters. As soon as this tension dissolves, viewers lose interest. However, one could recognize two conditions upon which a text would not exhibit violent signs.

First, not every sign would fall into a category. The volume

of possible sign-category would necessarily increase since each becomes unique and non-concretizing. Signs precede characters in
this sense. In *Frasier* it is the reverse: sign-potential is dependent upon the character producing the sign. Secondly, signs are
unevenly and non-selectively spread across bodies (of characters,
texts, spaces, etc.). Signs are singularities before they are categor-
ical representations. These conditions compose the two essential
states that a structure could take: stagnation or movement.

  The first condition relates to Simondon's individuation in which
intensities converge. The second is engendered by Derrida's play
in which signifiers endlessly appear without reaching a clear and
present signified. *Frasier* and its characters represent a structure
that is highly individuated with very little potential for play. Examples that tend towards play and do not exhibit violent oppositions
between signs are numerous, but an obvious one would be an am-
ateur video called "Rifle Burs" in which signs are spread across
all characters and bodies with varying techniques.[8] Hence, it must
be remembered that *Frasier* is merely one type of structure and is
not emblematic of all structures. The strength of hyphology rests
in its ability to objectively analyze structures without determining
universal limits of that structure.[9]

---

  8. "Rifle Burs," YouTube video, 3:22, accessed March 25, 2012, http://
www.youtube.com/watch?v=BrU_ef7DQgs.
  9. Deleuze's comment that all phenomenology is mere epiphenomenology
applies here (Deleuze, *Difference and Repetition*, 52). Deleuze's structuralism
of the virtual aimed at noumena rather than phenomena. The same could be
said about hyphology's relation to orthodox structuralism.

7

Conclusion

While working in an elementary school, a girl in third grade told me:
"It is not the back [of the dollar bill] that matters; only the front
does." At first glance this comment does not appear very disturbing,
but upon closer inspection, it reveals the profound distress we have
inherited from Derrida. Of course, I did not get the chance to ask
the young girl if she was explicitly making a commentary on the
impossibility of reaching a signified. Yet it is this emphasis on the
front (i.e. signifier) that one finds all throughout *Of Grammatology*.
It is this heritage of deconstruction that I have attempted to deal

with in this essay.

The broad scope of this essay is illustrative of its intention. To discuss Derrida we had to look at Barthes, to assess Barthes we turned toward Deleuze, to understand Deleuze we looked at Simondon, and so on. There is never an outside of the text because the periphery is always pushing in—this is something that every figure discussed in this essay would agree with. The form of this essay mirrors the form of structures. There is always something more that is necessary in order to clarify what is already there. Hyphology is an analysis of the trace; it is an *exploration*, in the sense that Barthes intended.

Deleuze's importance could not be stressed enough. He is introduced for the purposes of reinterpreting the role of repetition in structuralism. He helped reclaim meaning from deconstruction through a focus on a notion of meaning that is immanent and local. Hyphology is focused on three features of structures: trace, time, and meaning, which are elucidated by Derrida, Barthes, and Deleuze, respectively. Although each figure can be linked to one of these features, they all provide insights relevant to all three features. The goal of this essay has not been to overturn Derrida's critiques entirely, but to assimilate the accurate observations into a structuralism that can continue to function.

Hyphology seeks to enumerate the multiple methods that structures appropriate in order to produce meaning. In this way, it will never be completely explicated as a theoretical position. Each structure embodies a unique configuration of signs and a creative grammar. By recognizing this fact, hyphology is able to analyze the structure-in-itself, rather than forcing it into a pre-determined framework.[1]

In many ways, hyphology is *Gelassenheit*. Tracing the relation between these terms illuminates the manner in which they interact. So, the hyphologist cannot adopt a method that can be replicated for every structure. In each encounter, the differential geology of a structure must be palpitated to account for its particular context. Hyphology expands with every analysis and the only thing left to do is practice it.

I have committed to a reading of certain writers in order to come to my formulation of hyphology, but the world has always been pregnant with such a method. To explicate it in this essay, I have gone through Derrida and Barthes. However, if one were to combine Félix Guattari and Bernard Stiegler's conceptions of the machine, one would come to the same conclusion. This endeavor would take another essay that is of a similar length and cannot be discussed here in depth, but this suggestion is illustrative of the theoretical significance that hyphology holds for all areas of thought. There is a time for all ideas, which means that at a certain point in time, the conditions for an idea are over-determined. Hyphology follows this trajectory. In a time when post-structuralist thought has already passed but no new idea has surfaced to take its place, the need for a new method is most pressing. As the child of phenomenology, structuralism, and post-structuralism (all of which play a role in this essay), hyphology is a philosophical method for the twenty-first century.

---

1. Derrida, for instance, contrasts preformationism (i.e. structuralism) with epigenesis in order to underscore the crucial role played by time Derrida, *Writing and Difference*, 23.

8

Appendix: Chart of Signs from

Frasier

Bourgeois Signs	Proletarian Signs	Ambiguous Signs
Clothing (suits)	Clothing (plaid)	Daphne
Wine / "Sherry"	Beer	Jazz music
Seduction	Promiscuity	Bullying
Refined taste	Liberal taste	
Articulate speech	Emotions	
Telescope	Martin's chair	
Piano(s)	Physical labor	
African Art	Womanizing	
Bookcase	Mysticism	
Paintings	Sports	
Maris	Disability	
Frasier's	Crime	
balcony	Slang	
Apartment #1901	Animals	
Door size		
Wiping public chairs		
(Niles)		

Practices
(Modes of Performance)

Exhibition
Imitation
Monumentalization

Transcendental Signified

Class struggle

Bibliography

Barthes, Roland. *Course in General Linguistics*. Edited by Charles Bally and Albert Sechehaye. Translated by Roy Harris. Chicago, IL: Open Court, 2006.

———. *Elements of Semiology*. Translated by Annette Lavers and Colin Smith. New York, NY: Hill and Wang, 1977.

———. *Mythologies*. Translated by Annette Lavers. New York, NY: Hill and Wang, 1972.

———. *S/Z: An Essay*. Translated by Richard Miller. New York, NY: Hill and Wang, 1974.

———. *The Eiffel Tower and Other Mythologies*. Translated by Richard Howard. Berkeley: University of California Press, 1979.

———. *The Pleasure of the Text*. Translated by Richard Miller. New York, NY: Hill and Wang, 1975.

———. *Writing Degree Zero*. Translated by Annette Lavers and Colin Smith. New York, NY: Hill and Wang, 1977.

Company, Ford Motor. "Freedom for the Woman Who Owns a Ford." The Henry Ford. Accessed January 14, 2020. http://ophelia.sdsu.edu:8080/henryford_org/03-23-2014/exhibits/showroom/1908/ads.html.

Daylight, Russell. *What if Derrida was Wrong about Saussure?* Edinburgh: Edinburgh University Press, 2011.

Deleuze, Gilles. *Difference and Repetition*. Translated by Paul Patton. New York, NY: Columbia University Press, 1994.

Deleuze, Gilles. *Proust and Signs: The Complete Text.* Translated by Richard Howard. Minneapolis: University of Minnesota Press, 2000.

Deleuze, Gilles, and Félix Guattari. *A Thousand Plateaus: Capitalism and Schizophrenia.* Translated by Brian Massumi. Minneapolis: University of Minnesota Press, 2009.

Derrida, Jacques. *Margins of Philosophy.* Translated by Alan Bass. Chicago: University of Chicago Press, 1982.

———. *Of Grammatology.* Translated by Gayatari Chakravorty Spivak. Baltimore: Johns Hopkins University Press, 1997.

———. *Positions.* Translated by Alan Bass. Chicago: University of Chicago Press, 1981.

———. *Writing and Difference.* Translated by Alan Bass. Chicago: University of Chicago Press, 1978.

Frasier. National Broadcasting Company (NBC), 1993–2004.

Hughes, Joe. *Deleuze and the Genesis of Representation.* London: Continuum, 2008.

———. *Deleuze's Difference and Repetition: A Reader's Guide.* New York, NY: Continuum, 2009.

Kierkegaard, Søren. *Repetition.* Translated by Howard V. Hong and Edna H. Hong. Princeton, NJ: Princeton University Press, 1983.

Protevi, John. *Political Affect: Connecting the Social and the Somatic.* Minneapolis: University of Minnesota Press, 2009.

Simondon, Gilbert. "Translation: Chapter 1 of Simondon's *Psychic and Collective Individuation.*" Speculative Heresy. Accessed October 6, 2008. http : / / speculativeheresy . wordpre ss . com / 2008 / 10 / 06 / translation - chapter - 1 - of - simondons-psychic-and-collective-individuation/.

Smith, Daniel, and John Protevi. "Gilles Deleuze." The Stanford Encyclopedia of Philosophy. Accessed January 14, 2020. `http://plato.stanford.edu/archives/spr2013/entries/deleuze/`.

Woolf, Virginia. *To the Lighthouse*. New York, NY: Oxford University Press, 2008.

"Rifle Burs." YouTube video, 3:22. Accessed March 25, 2012. `http://www.youtube.com/watch?v=BrU_ef7DQgs`.

www.ingramcontent.com/pod-product-compliance
Lightning Source LLC
Chambersburg PA
CBHW071318220526
45468CB00001B/413